T0402027

4
Genome Dynamics and Stability

Series Editor: Dirk-Henner Lankenau

Available online at
SpringerLink.com

Transposons and the Dynamic Genome

Volume Editors: Dirk-Henner Lankenau, Jean-Nicolas Volff

With 36 Figures

 Springer

Series and Volume Editor:

Priv.-Doz. Dr. Dirk-Henner Lankenau
Hinterer Rindweg 21
68526 Ladenburg
Germany
e-mail: d.lankenau@t-online.de

Volume Editor:

Prof. Dr. Jean-Nicolas Volff
Ecole Normale Supérieure de Lyon
Institute de Génomique Fonctionnelle
46 alleé d'Italie
69364 Lyon Cedex 07
France
e-mail: Jean-Nicolas.Volff@ens-lyon.fr

Cover

The cover illustration depicts two key events of DNA repair: 1. The ribbon model shows the structure of the termini of two Rad50 coiled-coil domains, joined via two zinc hooks at a central zinc ion (sphere). The metal dependent joining of two Rad50 coiled-coils is a central step in the capture and repair of DNA double-strand breaks by the Rad50/Mre11/Nbs1 (MRN) damage sensor complex. 2. Immunolocalization of histone variant γ-H2Av in γ-irradiated nuclei of *Drosophila* germline cells. Fluorescent foci indicate one of the earliest known responses to DNA double-strand break formation and sites of DNA repair.
(provided by Karl-Peter Hopfner, Munich and Dirk-Henner Lankenau, Heidelberg)

ISSN 1861-3373   e-ISSN 1861-3381
ISBN 978-3-642-02004-9   e-ISBN 978-3-642-02005-6
DOI 10.1007/978-3-642-02005-6
Springer Dordrecht Heidelberg London New York

Library of Congress Control Number: 2009929233

Editor: Dr. Sabine Schwarz
Desk Editor: Ursula Gramm, Heidelberg
Cover figures: Prof. Karl-Peter Hopfner and Dr. Dirk-Henner Lankenau
Cover design: WMXDesign GmbH, Heidelberg
Typesetting and Production: le-tex publishing services GmbH, Leipzig

Printed on acid-free paper

Springer is part of Springer Science+Business Media (www.springer.com)

Preface

Sometimes my students and others have asked me: "what was first in evolution – retroviruses or retrotransposons?" Since Howard Temin proposed that retroviruses evolved from retrotransposons (Temin 1980; Temin et al. 1995) the other alternative that retroviruses emerged first and were the predecessors of LTR-retrotransposons has since been a controversial issue (Terzian et al., this BOOK). While DNA-transposons could not have existed in an ancestral RNA-world by definition, sure enough, some arguments definitely point towards a pre-DNA world scenario in which retroelements were the direct descendants of the earliest replicators representing the emergence of life. First, these replicators likely catalyzed their own or other's replication cycles via the catalytic properties of RNA molecules. After translation had emerged some replicators possibly encoded an RNA polymerase first. This later evolved into reverse transcriptase (RT), i.e. the most prominent key-factor at the transition into the DNA world. Simultaneously, replicators could also have encoded membrane protein-genes such as the *env* gene of recent DNA-proviruses. Membranes were likely present much earlier as prebiotic oily films that supported the evolution of a prebiotic-protometabolism (Dyson 1999; Griffiths 2007). However, how

these promiscuous communities of ancestral molecules and protocells inter-
acted, and how the exact branching chronology of earliest events in molecular evolution led to the emergence of replicators, membrane slicks, obcells
(Cavalier-Smith 2001) still remains a mystery. It still underscores Charles Darwin's statement cited top left, while Barbara McClintock's remark more than
100 years later (cited top right), represents the spirit for not giving up these
most fundamental topics.

One scenario is very likely: from the geochemically dominated times of
the early planet earth, prebiotic promiscuous communities including membranes, proto-peptides, metabolites, and replicators represented the ingredi-
ents of Darwin's *"wholly unknown process."* From these, we now think, life
emerged in conformity with a dual definition of life based on genetics and
metabolism.[1]

The platform for transposon-research is simple. Besides "genes," transposable elements evolved as indwelling entities within all cellular genomes.
Thereby, they exhibited both a parasitic as well as a symbiotic double-feature
that may date back to the very beginnings of life itself. Celebrating Charles
Darwin's bicentenary this year, we certainly do well to honor the fact that Darwin's concept of gemmules directly led to our present day term *"genes"* (Gould
2002; Lankenau 2007b). How pleased would Darwin have been to see this idea
brought onto the right track, e.g. through the works of Mendel, Weismann,
deVries, or McClintock. How pleased would he have been to know how close
we come today to his grand challenge: "The Origin of Species." Darwin, in fact
even came as close as he could to humanities deepest concern formulating his
famous statement:

*"It is often said that all the conditions for the first production of a living
organism are now present, which could ever have been present. But if (and
oh! what a big if!) we could conceive in some warm little pond, with all sorts of
ammonia and phosphoric salts, light, heat, electricity, &c., present, that a protein
compound was chemically formed ready to undergo still more complex changes,
at the present day such matter would be instantly devoured or absorbed, which
would not have been the case before living creatures were formed."* (Charles
Darwin 1871).

This statement also perfectly highlights our current technical hitches – but
some have been overcome, and transposable elements have their share in approaching the solution of the grand enigma. How pleased would Darwin have
been if he could have shared our modern insights into transposon-biology –
as we now understand some of the inner workings of transposon activities and

---

[1] Life is defined synergistically as the merging of replication and metabolism. H.J. Muller wrote: *It is
to define as alive any entities that have the properties of multiplication, variation and heredity* (Muller
1966). While metabolism supplies the monomers from which the replicators (i.e. genes or transposable
elements) are made, replicators alter the kinds of chemical reactions occurring in metabolism. Only
then can natural selection, acting on replicators, power the evolution of metabolism (Dyson 1999;
Maynard Smith and Szathmary 1997).

of analogous selfish genetic elements that triggered molecular, coevolutionary chases through sequence space and the emergence of driver systems resulting in "molecular peacock's tails" such as "autosome killer-chromosomes," "selfish sex chromosomes," and "genomic imprinting machineries." Despite his surmise that present day metabolism would devour or absorb all ancient metabolic systems, we now understand that a great deal of ancient bits of information survived inside the chromosomes of all organisms in the form of sequence relicts. A lot of these ancient molecular relicts belong to the stunning, endogenous survival machines that always represented the major engines of evolution since the times of the genetic takeover – in a sense they form the pillars of life, capable of shaping the evolution of genomes and opportunistically altering genome structure and dynamics: transposable elements and viruses as their extracellular satellites, that fill our world's oceans with an unimaginable number of 10^{31} entities, or else, 10^7 virions per ml of surface seawater (Bergh et al. 1989; Williamson et al., 2008).

In fact, life began as and is driven by an emergent self-organizing property. Transposable elements seem to have played a significant role as executors of Gould's/Eldgredge's Punctuated Equilibrium[2]. How are transposable elements defined and why are they important? Transposable elements are specific segments of genomic DNA or RNA that exhibit extraordinary recombinational versatility. Treating a transposable element as an individual biological entity, it is best defined as a *natural, endogenous, genetic **toolbox of recombination**. This entity also overlaps with a wider definition of the term gene.[3] A transposable element is typically flanked by non-coding, direct, or inverted repeat sequences of limited length (less than 2 kb) often with promoter- and recombinational functions. These repeats flank a central core sequence, which among few other genes encodes a transposase/integrase and/or reverse transcriptase (RT). Transposable elements are the universal components of living entities that appear to come closest in resembling the presumed earliest replicators (including autocatalytic ribozymes) at the seed crystal level of the origins of life. Stuart Kauffman realized that Darwinian theory must be expanded to recognize other sources and rules of order based on the internal numeric, genetic, and developmental constraints of organisms and on the structural limits and contingencies of physico-chemical laws (Kauffman 1993). While Kauffman's approach is a step toward a deep theory of homeostasis, it is smart to define

---

[2]Originally Stephen Gould's and Niels Eldredges' punctuated equilibrium theory holds that most phenotypic differences occur during speciation periods but that species embedded in stable environments are remarkable stable in phenotype thereafter (Eldredge and Gould 1972). Here, the expression "phenotypic stability" is extended beyond this definition that focused on biological species. The molecular structure of genomes exhibits an analogous platform of stable order. "Genes" and "transposable elements" are examples of such a stable platform of order with emergent self-organizing properties – see also: (Kauffman 1993).

[3]In a broad context, a gene is defined as any portion of chromosomal material that potentially lasts for enough generations to serve as a unit of natural selection (Dawkins 1976).

the starting point of life as the *catalytic closure*[4] of two elementary systems intrinsic to all forms of cellular life: (1) prebiotic protometabolism and (2) genetic inheritance[5] encompassing transposon-like replicators. Both (1) and (2) formed a duality at the emergence of life. As for Newton's second law of motion ($F = ma$) the couplet of terms metabolism and inheritance is defined in a circle; each (gene and biotic metabolism) requires the other. In fact, this circularity lay behind Poincaré's conception of fundamental laws as definitional conventions (Kauffman 1993). Further, the logical separation of the two is technical only and for argumentational, experimental purposes it is useful. On the primordial earth, ordered prebiotic proto-metabolism (Dyson 1999) likely congregated in the vicinity of geochemically formed membrane surfaces or within hemicells or obcells as Cavalier-Smith called them (Cavalier-Smith 2001; Griffiths 2007). Such earliest metabolically ordered environments perhaps were too dynamic to establish long chained replicators such as RNA. At present it appears more realistic to assume the origin and growth of long RNA molecules in sea ice (Trinks et al. 2005). Freeman Dyson unfolded a possible series of evolutionary steps establishing the modern genetic apparatus, with the evolutionary predecessors of transposable elements (i.e. replicators) at the heart of this process, establishing the modern genetic apparatus. Let us assume that the origin of life "took place" when a hemicell contained an ordered, homeostatically stable metabolic machinery (compare the similar ideas of Cavalier-Smith 2001). This system maintained itself in a stable homeostatic equilibrium. The major transition, establishing life was the integration of RNA as a self-reproducing cellular "parasite" but not yet performing a symbiotic genetic function for the hemicell. This transitional state must have been in place before the evolution of the elaborate translation apparatus linking the two systems could begin (Dyson 1999). The first replicators were not yet what we call transposable elements sensu stricto. They still had to evolve genes for proteins such as integrase and reverse transcriptase (RT). This transitional state of merging metabolism and replication represented the first of life's punctuated equilibria (Gould 2002) resulting in the inseparable affiliation of parasitic/symbiotic interactions of metabolites and replicators. The inseparable affiliation of symbiotic/parasitic features is the most typical characteristic of transposable elements active within modern genomes. After the genetic code and translation had been invented, and when the first retroelements evolved RT from some sort of RNA replicase, transposable elements (i.e. retroelements) triggered yet another punctuated equilibrium, i.e. the transition from the RNA world to an RNA/DNA world. Amazingly, the deep window into earth's most ancient past is still reflected by the vivid actions of transposable elements and viruses within all present-day genomes – it also includes the significant chimerical feature of parasitic versus symbiotic interdependencies. From time to time – typically, as evolution is

---

[4]*Catalytic closure* is defined as a system where every member of the autocatalytic set has at least one of the possible last steps in its formation catalyzed by some member of the set, e.g. peptides and RNA.

[5]See footnote 1

tinkering (Jacob 1977) – transposable element sequences that usually evolve under the laws of selfish and parasitic reproductive constraints became domesticated as useful integral parts of cellular genomes. One of the most forceful examples is the repeated domestication of sequence fragments from an endogenous provirus reprogramming human salivary and pancreatic salivary glands during primate evolution (Samuelson et al. 1990). The other prominent example of transposon domestication is the evolution of V(D)J recombination from the "RAG-transposon" crucial for the working of our immune system (Agrawal et al. 1998).

The above considerations force us to discern the historic rootage of transposable elements in geological deep time. The following chapters will serve sketching some of the enduring consequences of the emergence of transposable elements as inseparable constituents of modern genomes – as indwelling forces of species, populations and cells, recent and throughout evolution. The first two chapters establish key aspects of the significance of transposon dynamics as major engines of evolution on the level of genomes, populations, and species. The first chapter summarizes general theoretical approaches to transposon dynamics applicable to prokaryotes, as well as eukaryotes, with emphasis on the parasitic nature of transposable elements. Arnaud Le Rouzic and Pierre Capy point out that the evolution of a novel transposon insertion is similar to the dynamics of a single locus gene exposed to natural selection, mutations, and genetic drift. Different "alleles" can coexist at each insertion locus, e.g., a "void" allele without any insertion, a complete insertion, and multiple variants of deleted defective, inactivated alleles progressively accumulating through mutational erosion. Even though not mentioned in this context, the first chapter nicely approaches the NK model of Stuart Kauffman that forms the conceptual backbone of his grand opus the "Origins of Order" (Kauffman 1993, pp. 40–43). In the NK model N is the number of distinct genes in a haploid genome while K is the average number of other genes which epistatically influence the fitness contribution of each gene. Le Rouzic and Capy address the problem of a stable equilibrium. This, perhaps in the future promises to become congruent with Kauffman's prediction that many properties of the fitness-landscapes created with the NK model appear to be surprisingly robust and depend almost exclusively upon N and K alone (Kauffman 1993, p. 44). The second chapter merges historical aspects of transposable element dynamics at the infra- and transspecific populational level with modern approaches at the epigenetic level. While transposable elements were first discovered by Barbara McClintock in maize, Christina Vieira et al. focus and underscore the importance of *Drosophila* as a model organism in transposon research and populational studies.

The third chapter by Agnès Dettai and Jean-Nicolas Volff exemplifies the SINE[6] retroelements as a model system of real novel insertions of transposable

---

[6]Short interspersed nuclear elements (SINEs)

elements within variable chromosomal sites. SINES are shown as key examples for the powerful mode of evolutionary genome dynamics. Novel insertions not only create new fitness landscapes on which selection can act but if established within all germline genomes of a species they become powerful molecular morphological markers that are employed for cladistic analysis identifying unambiguous branching points in phylogenetic trees. This chapter truly represents the legacy of Willi Hennig's phylogenetic systematics (Hennig 1966; Hennig 1969) on a modern molecular platform. The chapter also lists a number of software tools making whole genome analysis feasible. Chapters 4 and 5 focus on transposable elements, and on the origin and regulation by means of double-stranded RNA and RNA interference (RNAi), another key-factor with evolutionary significance. While King Jordan and Wolfgang Miller review the control of transposable elements by regulatory RNAs and summarize general aspects of genome defense Christophe Terzian et al. in Chapter 5 present insights into the most interesting and the first example of an insect retrovirus, i.e. the endogenous *gypsy* retrotransposon of *Drosophila*. This retrovirus indeed represents an unmatched model system for multiple aspects of the biology of endogenous retroviruses as well as of an active retrotransposon. The *gypsy* provirus had been studied previously in connection with the host encoded Zn-finger protein Suppressor of Hairy Wing [Su(Hw)]. This protein turned out to be a chromatin insulator regulating chromatin boundaries and controlling enhancer-driven promoter activities. Its repetitive binding site within the *gypsy* provirus must have evolved within the *gypsy* retroelement by means of transposon evolution, perhaps in a quasispecies-like way. It is one of the most impressive examples demonstrating the emergence of the potential power of novel regulatory functions within host genomes (Gdula et al. 1996; Gerasimova and Corces 1998; Gerasimova et al. 1995). Terzian et al. (Chapter 5) advance our understanding and broaden our insights of *gypsy* driven by piRNA control mechanisms located within the heterochromatic *flamenco* locus. They further review recent findings as to the role of the envelope (Env) membrane protein serving as a model for retroviral horizontal and vertical genome transfer.

Another spectacular evolutionary example is presented in Chapter 6 by Walisko et al. It is the story of the revitalization of an ancient inactive DNA transposable element called *Sleeping Beauty*. It was reconstructed based on conserved genomic sequence-information only in the laboratory. The story is like Michael Crichton's Jurassic Park scenario, where dinosaurs were reconstructed from DNA in mosquito blood fossilized in amber. While Crichton's experiments were fiction, *Sleeping Beauty* is a real, reanimated "transposon-dinosaur." It existed for millions of years as an eroded, defective molecular fossil within a fish genome and was reactivated to study host-cell interactions in experimentally transfected human cells. Last but not least, the final chapter by Izsvák et al. describes the interactions of transposable elements with the cellular DNA repair machinery. Barbara McClintock first recognized the interdependence of chromosome breaks and transposition in her famous breakage-fusion-bridge cycle (McClintock 1992 (reprinted)). In the early 1990s Bill Engels and co-workers discovered the fundamental, prominent double-strand break repair mechanism they called Synthesis-Dependent Strand Annealing (SDSA) as the underlying molecular mechanism repairing P-transposable element-induced double-strand breaks. This mechanism of homologous recombination is now widely recognized and its role in genome dynamics is interwoven into many volume chapters of this book series. As regards content Chapter 7 therefore closes the cycle and links this fourth book volume of the series to the first volume integrating multiple aspects of genome integrity (Lankenau 2007a).

Altogether, this book gives insight and a future perspective regarding the significance of transposable elements as selfish molecular drivers and universal features of life that exhibit in the words of Burt and Trivers "a truly subterranean world of sociogenetic interactions usually hidden completely from sight" (Burt and Trivers, 2006).

I most cordially thank all chapter authors for contributing to this volume on genome dynamics and transposable elements. Most importantly, I am deeply grateful to all the referees whose names must be kept in anonymity. At least two for each chapter were involved in commenting, shaping, and struggling with the individual scripts – I really, greatly appreciate their efforts! I thank Jean Nicolas Volff for organizing the transposable element meeting at Wittenberg some time ago and helping to invite some of the authors. I also thank the editorial staff at Springer who have always been patient with the editors and authors alike and have provided much help. I especially thank the managing editor Sabine Schwarz at Springer Life Sciences (Heidelberg) and the desk editor Ursula Gramm (Springer, Heidelberg) for their enduring assistance. I would also like to mention that le-tex publishing services oHG, Leipzig did a good job in production editing and preparing the manuscripts for print.

Ladenburg, April 2009   Dirk-Henner Lankenau

References

Agrawal A, Eastman QM, Schatz DG (1998) Transposition mediated by RAG1 and RAG2 and its implications for the evolution of the immune system. Nature 394:744–751

Bergh O, Borsheim KY, Bratbak G, Heldal M (1989) High abundance of viruses found in aquatic environments. Nature 340:467–468

Burt A, Trivers R (2006) Genes in Conflict. The Belknap Press of Harvard University Press, Cambridge, Ma; London

Cavalier-Smith T (2001) Obcells as proto-organisms: membrane heredity, lithophosphory-lation, and the origins of the genetic code, the first cells, and photosynthesis. J Mol Evol 53:555–595

Dawkins R (1976) The selfish gene. Oxford University Press, Oxford

Dyson FJ (1999) Origins of life, Rev. edn. Cambridge University Press, Cambridge, U.K.; New York

Eldredge N, Gould SJ (1972) Punctuated equilibria: An alternative to phyletic gradualism. In: Schopf TJM (ed) Models in palaeobiology. Freeman Cooper, San Francisco, pp 82–115

Gdula DA, Gerasimova TI, Corces VG (1996) Genetic and molecular analysis of the gypsy chromatin insulator of *Drosophila*. Proc. Natl. Acad. Sci. U S A 93:9378–9383

Gerasimova TI, Corces VG (1998) Polycomb and Trithorax group proteins mediate the function of a chromatin insulator. Cell 92:511–521

Gerasimova TI, Gdula DA, Gerasimov DV, Simonova O, Corces VG (1995) A *Drosophila* protein that imparts directionality on a chromatin insulator is an enhancer of position-effect variegation. Cell 82:587–597

Gould SJ (2002) The structure of evolutionary theory. Belknap Press of Harvard University Press, Cambridge, Mass., USA

Griffiths G (2007) Cell evolution and the problem of membrane topology. Nat Rev Mol Cell Biol 8:1018–1024

Hennig W (1966) Phylogenetic Systematics. University of Illinois Press, Illinois, USA

Hennig W (1969) Die Stammesgeschichte der Insekten. Vlg. Waldemar Kramer, Frankfurt

Jacob F (1977) Evolution and tinkering. Science 196:1161–1166

Kauffman SA (1993) The origins of order: self organization and selection in evolution. Oxford University Press, New York

Lankenau D-H (2007a) Genome integrity: Facets and perspectives. Springer, Berlin Heidelberg New York

Lankenau D-H (2007b) The legacy of the germ line – maintaining sex and life in metazoans: Cognitive roots of the concept of hierarchical selection. In: Egel R, Lankenau D-H (eds) Recombination and meiosis – Models, means and evolution, vol 3. Springer, Berlin Heidelberg New York, pp 289–339

Maynard Smith J, Szathmary E (1997) The major transitions in evolution. Oxford University Press, Oxford

McClintock B (1992 (reprinted)) Chromosome organization and genetic expression. In: Fedoroff N, Botstein D (eds) The dynamic genome: Barbara McClintock's ideas in the century of genetics. Cold Spring Harbor Laboratory Press, Cold Spring Harbor, N.Y., pp 73–107

Muller HJ (1966) The gene material as the initiator and organizing basis of life. Am Nat 100:493–517

Samuelson LC, Wiebauer K, Snow CM, Meisler MH (1990) Retroviral and pseudogene insertion sites reveal the lineage of human salivary and pancreatic amylase genes from a single gene during primate evolution. Mol Cell Biol 10:2513–2520

Temin HM (1980) Origin of retroviruses from cellular moveable elements. In: Cell, vol 21, pp 599–600

Temin HM, Cooper GM, Temin RG, Sugden B (1995) The DNA provirus: Howard Temin's scientific legacy. ASM Press, Washington, D.C.

Trinks H, Schröder W, Biebricher CK (2005) Ice and the origin of life. Orig Life Evol Biosph 35:429–445

Williamson SJ et al. (2008) The Sorcerer II Global Ocean Sampling Expedition: metagenomic characterization of viruses within aquatic microbial samples. PLoS ONE 3:e1456

Contents

Interactions of Transposons with the Cellular DNA Repair Machinery

Genome Dyn Stab (4)
D.-H. Lankenau, J.-N. Volff: Transposons and the Dynamic Genome
DOI 10.1007/7050_017/Published online: 15 July 2006
© Springer-Verlag Berlin Heidelberg 2006

Theoretical Approaches to the Dynamics of Transposable Elements in Genomes, Populations, and Species

Arnaud Le Rouzic[1,2] · Pierre Capy[1] (✉)

[1]Laboratoire Évolution, Génétique et Spéciation (CNRS), Avenue de la Terrasse,
Bâtiment 13, 91198 Gif sur Yvette, France
pierre.capy@legs.cnrs-gif.fr

[2]*Present address:*
Linnaeus Centre for Bioinformatics, Uppsala Universitet, 75124 Uppsala, Sweden

Abstract Transposable elements are major components of both prokaryotic and eukaryotic genomes. They are generally considered as "selfish DNA" sequences able to invade the chromosomes of a species in a parasitic way, leading to a plethora of mutations such as insertions, deletions, inversions, translocations and complex rearrangements. They are frequently deleterious, but sometimes provide a source of genetic diversity. Numerous population genetics models have been proposed to describe more precisely the dynamics of these complex genomic components, and despite a wide diversity among transposable elements and their hosts, the colonization process appears to be roughly predictable. In this paper, we aim to describe and comment on some of the theoretical studies, and attempt to define the "life cycle" of these genomic nomads. We further raise some new issues about the impact of moving sequences in the evolution and the structure of genomes.

1
Introduction

Transposable Elements (TEs) seem to be an outstanding example of evolutionary success. They are present in almost all known living species, from eubacteria and archaebacteria to the multicellular organisms. They show a huge genetic and functional diversity, and they seem to have explored during the evolution process, the most relevant ways possible to duplicate and maintain themselves in the genome of their "host". The persistence of TEs in the genome, sometimes in spite of significant deleterious effects, is generally attributed to their amplification ability. This is the basis of the "selfish DNA" theory (Orgel and Crick 1980; Doolittle and Sapienza 1980; Hickey 1982).

Selfish DNA sequences appear to be submitted to several antagonistic multi-level forces, driving them along various evolutionary pathways. These depend on multiple factors, such as the biology of the host species, the features of the TE family, or simply chance. TE dynamics can be quite complex such that further analysis rests on mathematical models of population genetics. At the molecular level, the more efficient the transposition process, the more likely the colonization of the genome will be. However, if the elements are deleterious for the host, individuals carrying too many copies will be eliminated through natural selection. Evolution of genomes would also certainly lead to the appearance of systems controlling or regulating replication, and elements are likely to evolve towards a way of bypassing such systems. Recurrent genomic mutations lead to partial or complete deletions or inactivations of TE copies, while some elements or fragments of elements may remain integrated in the genome and participate in an adaptive function of the organism. In this chapter, we propose to review the interactions existing between a genome and such internal parasites from a population genetics point of view. These interactions can change radically between the several successive stages of the invasion, from the active colonization of the genome by elements, to the probable loss of the transposition activity.

2
Genome Colonization

Theoretical studies of TE dynamics are generally challenged by the complexity of the process (see Charlesworth et al. 1994; Le Rouzic and Deceliere 2005 for review). The evolution of each TE insertion is actually similar to the dynamics of a single locus gene exposed to natural selection, mutations, and genetic drift. Different "alleles" can coexist at each insertion locus (e.g., a "void" allele without any insertion, a complete insertion, and multiple deleted, defective, inactivated alleles progressively appearing through mutations), and each of them might have different transposition rates and different impacts on the fitness in heterozygous or homozygous states. Depending on the stage in the invasion and on the features of the element, several insertions, often a few dozens and sometimes much more, have to be considered simultaneously. Finally, the total number of insertion sites is thought to vary, each transposition event leading to a new insertion locus.

2.1
Copy Number Dynamics

Except for complex computer simulations, modelling such a system must be achieved through approximations. For instance, the initial invasion of the element in a void population can be modelled in the same way as segregation distortion, considering only one insertion locus (Hickey 1982). However, this approach does not give us the opportunity to explore the subsequent steps of the invasion, when TEs accumulate in the genome, and it therefore becomes necessary to consider average copy numbers. Charlesworth and Charlesworth (1983), for example, proposed to describe the variation of the average copy number $\bar{n}$ by $\Delta\bar{n} \simeq \bar{n} \cdot (u - v)$, where u is the transposition rate and v the deletion rate. This transposition (respectively deletion) rate corresponds to the mean number of transposition (or deletion) events for one copy in one generation. "Transposition" and "deletion" have to be understood here as generic terms aiming to include multiple kinds of molecular events, since only the resulting state is considered: a transposition (or, more precisely, a duplication) event leads to the appearance of a copy at a new insertion site, while a deletion results in the lost of a copy from its original insertion site[1]. This model is supposed to be approximately universal (i.e., all known TEs can fit with this model provided u and v are set accurately). If $u > v$, the element is able to invade, and the copy number increasing is exponential (Fig. 1). However, such dynamics do not appear realistic, since an infinite multiplication of a TE in a genome probably leads to its destruction. Two main evolutionary forces are supposed to be able to counterbalance this invasion: transposition regulation and natural selection (Fig. 2).

Transposition regulation consists in a decrease of the transposition rate during the invasion[2]. It can be roughly modelled by a transposition rate (i.e., duplication rate) $u_{\bar{n}}$ which is dependent on the mean copy number in the population $\bar{n}$: the higher the copy number, the lower the transposition rate. When the transposition rate $u_{\bar{n}}$ is equivalent to the deletion rate v, then $\Delta\bar{n} = 0$ and an equilibrium state is achieved (Fig. 1). However, this equilibrium situation supposes that $u = v$, which is generally not verified in natural populations, where transposition rates are usually at least one order of magnitude higher than the deletion rates (Nuzhdin and Mackay 1995; Suh et al. 1995; Maside et al. 2000). It, therefore, appears unlikely that transposition regulation is the only evolutionary force implied in TE copy number control.

Due to their activity, TEs represent a potential source of a large spectrum of mutations and chromosomal rearrangements. These mutations have been shown to be generally deleterious (Eanes et al. 1988; Mackay et al. 1992; Charlesworth 1996; Houle and Nuzhdin 2004), and natural selection is

---

[1] Class I elements (retrotransposons) transpose by a replicative mechanism, often referred as "copy and paste"; they can, however, be lost – or duplicated (Lankenau et al. 1994) – through other mechanisms, such as recombination between the terminal repeats of LTR retrotransposons (Vitte and Panaud 2003), or by synthesis dependant strand annealing (SDSA) (Lankenau and Gloor 1998). On the contrary, class II transposons move through a "cut and paste" mechanism; they are excised from the donnor site and reinserted at a new locus. They are, however, frequently duplicated through a homologous template dependant process (Brookfield 1995). Even if these mechanisms are not related, the overall dynamics of a TE family can be described by a transposition rate and a deletion rate, and interestingly, the order of magnitude of these parameters do not appear to be very different across TE classes (Hua-Van et al. 2005).

[2] This phenomenon has been described for many elements in several species (Labrador and Corces 1997). It is particulary well documented in intensively studied systems, such as P element in *Drosophila melanogaster* and its *KP* repressor (Jackson et al. 1988; Simmsons et al. 1990; Corish et al. 1996).

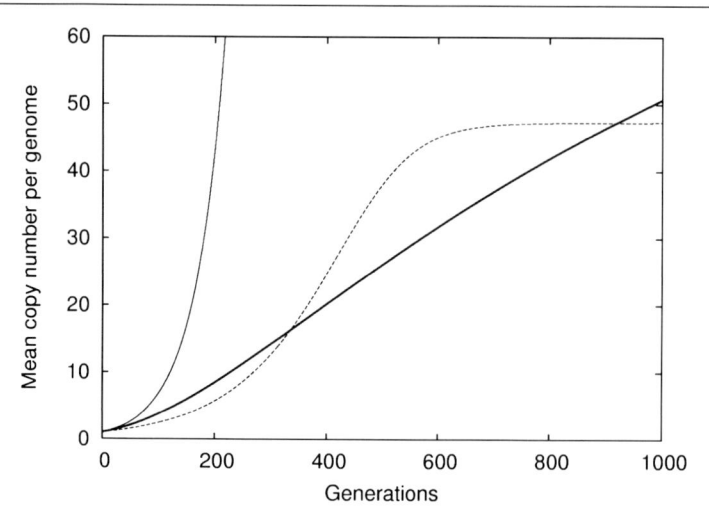

Fig. 1 Basic transposable element dynamics. If the transposition rate (frequency of a duplication event per copy and per generation) as well as the deletion rate (probability for a copy of being lost by various processes – see text) are constant, without any selection, the copy number increases exponentially ($\Delta n = n \cdot (u - v)$, with $u = 0.02$ and $v = 0.001$, *thin continuous line*). This probably does not correspond to a realistic situation, and several hypotheses have been proposed to explain the limitation of TE amplification (Charlesworth and Charlesworth 1983): (i) a regulation system, which supposes that the transposition rate decreases with the copy number: $\Delta n = n \cdot (u_n - v)$, with $u_n = u/(1 + k \cdot n)$, k being a factor that quantifies the intensity of regulation (here, $k = 0.2$, *thick line*); (ii) natural selection that eliminates, in each generation, a part of the insertions from the genome; $\Delta n = n \cdot (u - v - \partial \log w_n/\partial n)$. The *dotted line* represents the dynamics of such a system, with $w_n = 1 - s \cdot n$ (additive effects of insertions), and $s = -0.01$ (i.e., each insertion decreases the fitness by 1%)

also likely to restrain the TE proliferation. In a polymorphic population, the individuals carrying the lower number of copies are more likely to reproduce, leading to a slight decrease, each generation, in the mean copy number. Charlesworth and Charlesworth (1983) proposed to model this process by $\Delta \bar{n} = \bar{n} \cdot (u - v - s_{\bar{n}})$, where $s_{\bar{n}} = |\partial \log w_{\bar{n}}/\partial \bar{n}|$, w_n representing the fitness of an individual carrying n copies (and $w_{\bar{n}}$ being the fitness of a virtual individual having the average number of copies $\bar{n}$, which is reasonably close to the average fitness of the population). This model does not always lead to a stable equilibrium (Fig. 1), depending on the shape of the fitness curve w_n (Fig. 3).

The two processes (i.e., regulation and selection) are not mutually exclusive, and one can easily imagine that the TE amplification can be subject to both of them. Well-known TE families, such as P element in *Drosophila*, indeed appear to be both regulated (Lemaitre et al. 1993; Coen et al. 1994) and selected against (Snyder and Doolittle 1988; Eanes et al. 1988). A sim-

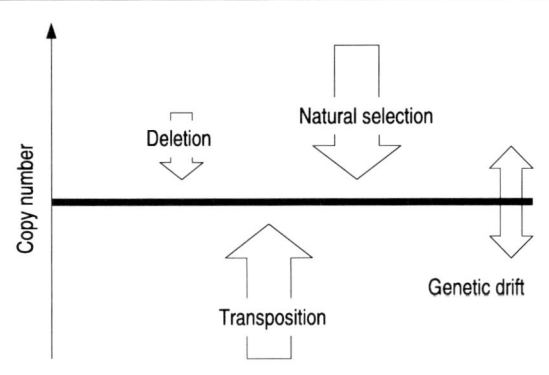

Fig. 2 Simple representation of the different evolutionary forces implied in the dynamics of TE copy number in the genome of a species. Transposition (or, more exactly, duplication) will increase the average copy number, while various kinds of transposition-related or unrelated deletions or excisions will eliminate copies from the genome. If the insertions are deleterious, the individuals carrying fewer copies will reproduce better than the others, and natural selection will decrease the mean copy number in the population. Several processes can be involved in this fitness loss: direct effect of insertions in genes or regulatory regions, repetitions leading to deleterious ectopic recombinations, or straight deleterious effect of the transposition activity (Nuzhdin 1999). Finally, in small populations, random genetic drift can shift the copy number below or above the expected value. At the beginning of the invasion process, the transposition rate is probably high, and the genomic copy number increases. A further equilibrium state can be achieved when increasing and decreasing forces are balanced; a decay in the transposition rate (recurrent mutations of active copies, transposition regulation …) or an intensification of the selective strengths can lead to this situation

ple model that combines both natural selection and transposition regulation shows that the effects of both evolutionary forces are cumulative (Fig. 4): if the transposition regulation is too weak to induce a realistic stabilization of the copy number, and if the selection strength alone is not sufficient to lead to an equilibrium (even if the fitness function does not match the conditions detailed in Fig. 3), then a perfectly realistic equilibrium copy number can be achieved when both control mechanisms overlap.

2.2
The Birth of a New TE Invasion

All these models describe the colonization of a TE family as a deterministic process. The spread of a TE in a population, and the progressive increase in the copy number does indeed appear as a predictable mechanism (e.g., Biémont 1994), provided the population size is large, thus limiting the influence of genetic drift (for the role of genetic drift in TE dynamics, see Brookfield and Badge 1997). However, regardless of the population size, an element cannot escape from randomness at the beginning of its invasion.

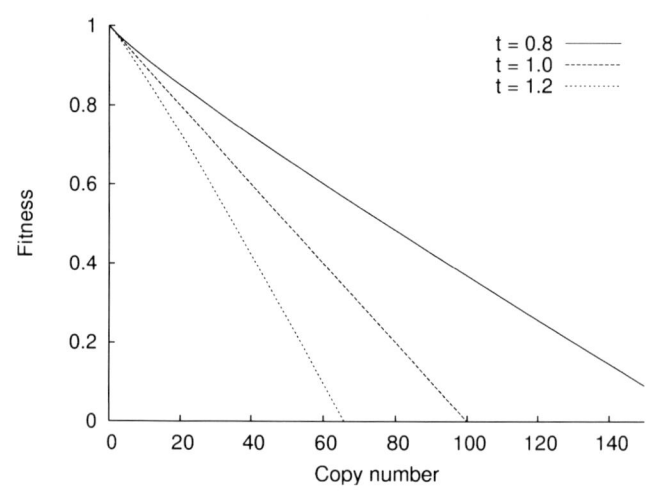

Fig. 3 The existence of a potential equilibrium state depends on the shape of the fitness curve (Charlesworth and Charlesworth 1983). The accumulation of TEs is supposed to be deleterious, and the fitness of an individual depends on the number of copies carried by its genome: the higher the copy number, the lower the fitness. However, an equilibrium can be achieved only if the fitness function is log-concave, i.e., if $\partial \log w_n / \partial n > 0$. The graph presents the shape of three different fitness functions, all based on the formula $w_n = 1 - s \cdot n^t$, which has been often used because its shape depends only on the parameter t: each insertion decreases the fitness by the same value ("additive model" with $t = 1$, *thick dotted line*), the absolute effect of insertion decreases during the invasion ($t = 0.8$, *continuous line*), or each new insertion is more deleterious than the previous ones ("multiplicative model", $t = 1.2$, *thin dotted line*). These different selection models may correspond to different mechanisms known to be related to TE-mediated mutations (Nuzhdin 1999). If the main cause of the deleterious effects of TEs relies in insertion effects (e.g., disruption of coding or regulatory sequences), the linear model could be likely. On the other hand, if the major part of the TE-induced genetic load correspond to chromosomal abnormalities due to ectopic recombinations between TE copies, the multiplicative model could be more appropriate, since the frequency of recombinations probably increases with the square of the copy number (Langley et al. 1988). The respective weights of these different factors are still poorly known (see Le Rouzic and Deceliere 2005 for review)

Each new element that colonizes the genome of a species derives from a closely related TE sequence coming from the same genome or from the genome of another species. Genomes are full of inactive or deleted TE copies, which can potentially recombine and generate a new, functional TE sequence. However, most TE invasions seem to be related to interspecific horizontal transfers (HTs), which remain anecdotal for eukaryotic "standard" genes (Davis and Wurdack 2004; Kurland et al. 2003), but much more frequent in TE evolution. Indeed, TEs are generally thought to show an amazing ability to "jump" between species (Kidwell 1992), whatever the phylogenetic distances between them (closely related *Drosophila*, Silva et al. 2004; Sanchez-Gracia et al. 2005, or different lineages of vertebrates, Leaver 2001).

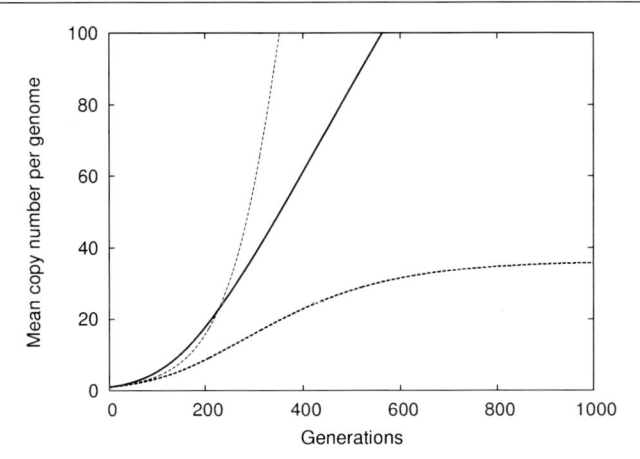

Fig. 4 The achievement of a realistic equilibrium depends on the strength of selection and regulation. If regulation or natural selection are too weak, no realistic equilibria can be expected (see also Fig. 1). On this figure, the *thin dotted line* represents a situation where the selection strength is low ($w_n = 1 - s \cdot n$ with $s = -0.005$) and the *thick continuous line* a situation where the regulation is weak ($u_n = u/(1 + k \cdot n)$ with $k = 0.05$, see Fig. 1 for the meaning of k). In both cases, the transposition rate is $u = 0.02$ and the deletion rate $v = 0.001$. The expected equilibria are achieved with very high, probably unrealistic, copy numbers. However, if this weak selection and regulation are combined (*thick dotted line*), the equilibrium copy number drops to fewer than 40 copies

In any case, when a TE arrives in an uncolonized species, its initial spread depends on its transposition rate, its selective impact on the new host species, and genetic drift (Fig. 5). Some specificities of the TE biology (such as the tissue or stage during development where transposition occurs, before, during or after the meiotic divisions[3]) may also alter the probability of fixation. The conditions leading to an effective invasion of an element appear to be rather complex: in the first stage of the colonization, the transposition rate has to be moderately high, but the maintenance of such an "aggressive" behaviour is likely to lead to an irreversible accumulation of deleterious mutations (Le Rouzic and Capy 2005). A theoretically "optimal" TE should, therefore, have a sophisticated "parasitic strategy", including a decrease of the transposition rate during the colonization process. The initial stage of high transposition can correspond to known "transposition bursts", that increase significantly the genomic copy number of one TE family[4] – and probably decrease the fitness of their hosts – in a few generations (Gerasimova et al. 1984; Biémont et al. 2003). The well known "hybrid dysgenesis", described in various *Drosophila* species for a couple of TE families (Bregliano and Kidwell 1983; Bucheton 1990; Vieira et al. 1998) might thus play a relevant role in TE

---

[3] For instance, early transposition events may lead to mutational clusters (Woodruff et al. 2004).

[4] or perhaps several families at the same time (Petrov et al. 1995).

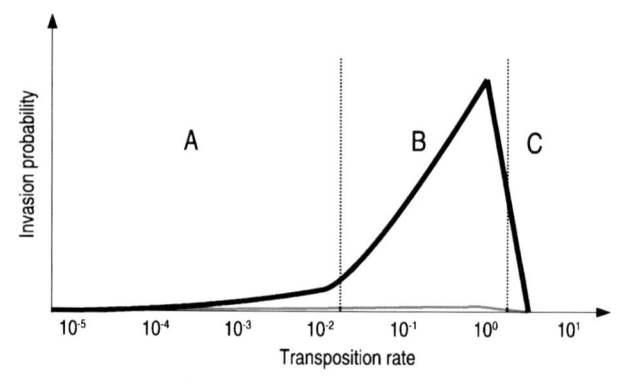

Fig. 5 The invasion capability of a TE family directly depends on its initial transposition rate. The figure represents the maintenance probability of a TE after 100 (*black line*) and 1000 (*grey line*) generations (from Le Rouzic and Capy 2005). One initial copy (simulating a horizontal transfer event) is introduced into a "naive" population. If the transposition rate is too low (**A**), the element is almost always lost through genetic drift and selection. If the transposition rate is very high (**C**), the spread of the TEs in the germline genome of single individuals is faster than their spread in the population: the fitness of the carriers of the element decreases and the element is lost. Finally, only a narrow range of moderately high transposition rates (**B**) allows an efficient invasion. The maximal invasion frequency depends on the selective coefficient and on the population size; it is generally less than 0.5 (i.e., the loss the the newly introduced TE remains the most frequent scenario). In any case, if this efficient transposition rate is maintained for a long time, TEs are likely to amplify, until they are responsible for a very high genetic load, leading to the extinction of the population in less than 1000 generations (*grey line*). Transposition regulation therefore appears as a necessary stage in the life cycle of a TE family

dynamics. Although complex, this "battle plan" might have been used by numerous TEs.

Self-regulation of TEs is certainly a representative example of a feature for which natural selection at the population level and intra-genomic selection are contradictory, and the resulting evolution appears to be hard to predict, since the occurence of self-regulation, although theoretically unlikely (Charlesworth and Langley 1986), seem to be nonetheless widespread (Labrador and Corces 1997), and some regulation mechanisms have been studied very intensively[5]. The decrease in the transposition frequency of an entire TE family after a high initial transpositional activity is indeed advantageous not only for the element, but also for the host. Regulation is often split into mechanisms due to the TE itself (self-regulation) and those due to host genes and/or epigenetic factors, but the particular components of the regulation system coming respectively from the host and from the element cannot

---

[5] For example, the *P* element in *Drosophila* and its regulatory element named *KP* have been analyzed at the molecular level (Jackson et al. 1989; Engels 1989; Rio 1991; Gloor et al. 1993; Andrews and Gloor 1995; Corish et al. 1996; Witherspoon 1999).

be generally determined. In fact, TEs are included in the genome, and their respective interests sometimes overlap. Some evolutionary constrains might also take place. For instance, transposition promoting selfish DNA invasion is required only in the germ cells; a high somatic transposition frequency is probably deleterious both for the host and the element. The regulation system is then adaptive for both entities, and the genomic conflict resides only in the control of the regulation process, and not in its existence. Regulation is therefore probably relevant to several interacting factors, such as selection on the colonization efficiency of TEs, fortuitous limiting mechanisms, and co-evolution between the TE and its host genome. Understanding the respective evolutionary impact of each of them actually presents a serious challenge.

Finally, the selective pressures applied to TE sequences are likely to be modified during the invasion process. The features necessary to colonize a population after a HT are certainly different from what is required for a long-term maintenance in the genome. Most known TE sequences seem to have experienced several effective transfers (Sanchez-Gracia et al. 2005), and TE families able to achieve successful HTs are certainly more likely to spread among the genomes of living organisms. HTs therefore probably play an extensive role in TE evolution (Lampe et al. 2003), and some widespread TE families could have maintained this ability, even if they are less efficient in further invasion steps. However, the HT rate of some TE families, such as LINE elements, appears to be very small (Burke et al. 1998), even though LINE elements are one of the most successful families in the genomes of vertebrates and many other species (Boissinot et al. 2000; Weiner 2002). Interspecific jumps do, therefore, not appear to be required for TE "survival".

3
TE – Genome Coevolution

3.1
Towards a Stable Equilibrium?

Despite a few exceptions (Ohta 1986; Quesneville and Anxolabéhère 2001), almost all TE dynamics models suppose that, after its initial invasion stage, the TE family reaches a stable equilibrium. This criterion has even been used as an argument to reject some "unrealistic" models (see for instance the models by Brookfield 1982, and by Charlesworth 1991), which do not lead to realistic stable states. However, experimental evidence about the persistence of a dynamic equilibrium state remain weak: laboratory experiments cannot be long enough to explore long-term evolution (e.g., Anxolabéhère et al. 1987; Montchamp-Moreau 1990; Biémont 1994), and complete sequences only provide a snapshot of the state of the genome at a given time. Theoretical studies have shown than the time necessary to reach an equilibrium state can be long

(Tsitrone et al. 1999), and any external event, such as demographic, environmental, or genomic disturbances, or even genetic drift, are likely to prevent the population from attaining this equilibrium.

Formally, the stability of the equilibrium state is based on the reversible nature of the mechanisms involved in this process. For instance, if the transposition rate decreases while the copy number increases, leading to a transposition – deletion equilibrium, then the transposition rate must grow in the same way if the copy number falls accidentally. The same kind of symmetry is also needed for the maintenance of an equilibrium based on natural selection. In fact, any small disturbance of the equilibrium state has to be exactly compensated by opposing selective forces (Fig. 2).

However, the real stabilization mechanisms are probably not so straightforward. On one hand, most of the regulation processes do not appear to be reversible. Some of them, such as repeat-induced point mutations (Hood et al. 2005), lead to the definitive destruction of the TE sequences, while others (RNAi mechanisms, for instance) probably persist even if some copies of the same family are eliminated. On the other hand, natural selection will tend to eliminate the most deleterious insertions, and the average insertion effect is certainly not constant over time. Finally, as for every genomic sequence, TEs are likely to accumulate mutations that will neutralize their transpositional activity. Some mutant copies might become non-autonomous elements, still able to transpose by parasiting the transposition machinery produced by autonomous copies, and thus probably decreasing the general transposition rate. All these phenomena, breaking the symmetry of the stabilization process, are likely to occur as soon as the system has reached its equilibrium point (or even before), preventing the maintenance of a constant copy number in the genome. The unlikeliness of the equilibrium state has also been confirmed by several theoretical models where mutant copies can appear (Kaplan et al. 1985), or where the selective effect of insertions are allowed to vary (Charlesworth 1991).

Most of the mechanisms that are expected to disrupt the equilibrium stage appear to lead to a decrease in the copy number of autonomous elements. The maximum amount of active TE sequences is thus likely to be reached immediately after the initial invasion. A short equilibrium (or pseudo-equilibrium) stage period can then occur, followed by a slow decay of the active TE content because of natural selection and spontaneous mutations and deletions (Fig. 6). This long-term dynamic probably depends not only on the features of the TE (e.g., transposition), but may also be influenced by the characteristics of the host (its ability to eliminate degenerated sequences, for instance, which seems to vary even between closely related species, Petrov and Hartl 1998) and by complex host-TE relationships (such as regulation processes).

Depending on the speed of the various stages of the invasion, the general dynamics can adopt different forms. If the mutation rate is low, or if the selection against TEs is weak, then the slope of the decay can be so slight such

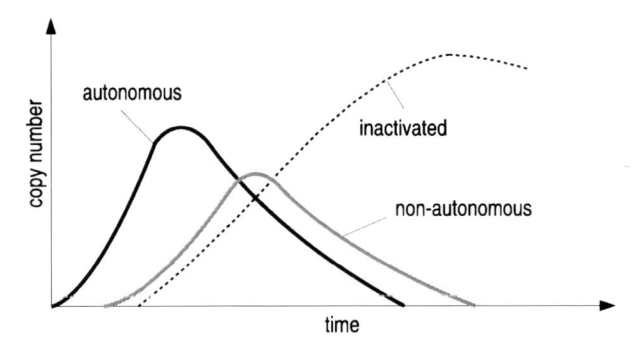

Fig. 6 Putative evolution of a TE family in a genome. After a rapid invasion stage of an active, autonomous element (*thick black line*), both reversible and non-reversible mechanisms will limit the transposition rate and the spread of new copies: the total copy number stabilizes, and then decreases, the copies being progressively eliminated by natural selection and by recurrent mutations. Mutant non-autonomous copies (*grey line*) can eventually take advantage of the remaining autonomous copies to multiply. Finally, the only TE-derived sequences persisting in the genome are inactivated elements (*dotted line*), which will be slowly eliminated and fragmented

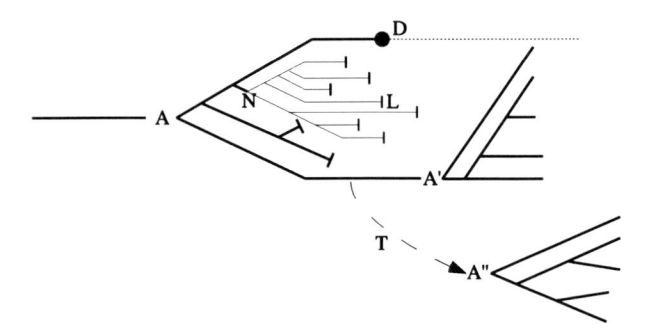

Fig. 7 Representation of the life cycle of a TE family. After its arrival in a new species, a single active TE copy (*thick line*) has to amplify itself (A), otherwise it will be rapidly eliminated by natural selection and genetic drift. The copy number then increases in the genome, and some mutations are likely to occur in these functional elements. Some of them (N) can lead to the appearance of non-autonomous copies (*thin lines*), which are able to amplify themselves provided complete, autonomous copies are present in the same genome. Some other copies may bring an adaptive feature to the host, so that they can be domesticated (D) and fixed, even if they lose their transpositional activity. But finally, the activity of the family stops and active elements are progressively lost (L), due to deletions and mutations, and perhaps because of a decrease in the transposition rate as a result of the multiplication of non-autonomous "selfish" elements. However, a few active autonomous copies might escape from this general decay, and can then initialize a new invasion process in the same species (A') or in another species (A") through a horizontal transfer (T). Even if the decay of the whole family after the colonization stage appears to be a determinist process included in the "life cycle" of a TE family (Kidwell and Lisch 2001), the accidental survival of copies active enough to start a new invasion cycle could be considered as a usual way of maintenance for TE sequences

that this phase may look like a pseudo-equilibrium situation. On the contrary, high mutation rates are likely to lead to the loss of any temporary equilibrium stage. Accidental events can also disturb this dynamic. For instance, an autonomous active copy can "survive" the decay (e.g. by means of concerted evolution based on homologous recombination mechanisms), and originate a new invasion cycle (Le Rouzic and Capy, unpublished). Other elements, inserted by chance in a locus where they are responsible for an adaptive feature to their host, may be fixed in the genome through molecular domestication (Miller et al. 1999). The potential occurrence of these events leads to the definition of several long-term evolution scenarios, which probably correspond to the wide diversity described among the insertion patterns of various TE families (Fig. 7).

3.2
Life Cycle of a TE Sequence

Numerous recent TE amplifications have been described in several organisms, such as *P* (Anxolabéhère et al. 1988; Engels 1997), *I* (Bonnivard et al. 2000), and *hobo* (Kidwell 1983) elements in *Drosophila melanogaster* or various TE families in plants (San Miguel et al. 1996; Feschotte and Mouchès 2000). Such new invasions are relatively frequent, but a general increase of the genome size is not what is usually reported (Petrov 2001); these invasions must be at least partially compensated by the loss of TE families. Indeed, numerous TEs seem to have disappeared from the genomes, and only non-functional copies can be identified, which are sometimes so divergent that they can be evidenced only through complex algorithms (Quesneville et al. 2003). This pattern suggests that there is a continuous flow of TEs in the genome, where they amplify before their activity ceases and they slowly become eliminated (Fig. 7). The long-term maintenance of a TE family in a large spectrum of species therefore seems to rely on accidental events, such as horizontal transfers, or the survival of an active element from the decay.

One of the main conceptual obstacles to our understanding of TE evolution is the multiplicity of the selection levels. From the host's point of view, at the population genetics time scale, TEs are unambiguously deleterious, and any TE invasion will substantially increase the genetic load carried by a population. However, on a larger time scale, TE mobility represents an outstanding source of genetic diversity (Mackay 1985; Kazazian 2000). More and more examples of TE domestications have been documented (e.g. several DNA-binding factors, Aravind 2000; Roussigne et al. 2003, the telomere elongation system in *Drosophila melanogaster*, Pardue and DeBaryshe 2003, or the V(D)J somatic recombination system in vertebrates Agrawal et al. 1998), a global survey of regulatory sequences shows that a significant number of them is derived from TE insertions in humans (Jordan et al. 2003) and in *Caenorhabditis elegans* (Ganko et al. 2001, 2003). TEs, thus, appear to be both deleterious and potentially adaptive (Capy et al. 2000; Brookfield 2003). The adaptive value of domestic TEs are most excellent examples of what has now been known as *exaptations*[6] in evolution (Gould 2002; Brosius 2005).

However, TEs are also submitted to another form of selection, i.e., intra-genomic selection (Snyder and Doolittle 1988). The different TE copies from the same family (or even from different families) are probably competing in the genome for various resources (e.g., the transposition machinery, some host factors, and probably other less easily accessible values such as the total genetic load that can be supported by thc host). Some important TE features, such as the regulation mechanisms, are likely to be affected by such a process.

Are transposable elements selfish, aggressive parasites, precisely optimized for taking advantage of their host? Selective pressures on TE sequences appear to be a mix between short-term (maintenance in generation after generation) and long-term selection (only TE families able to escape decay can be successful in evolutionary terms), between inter-individual and intra-genomic selection, and between several different successive genomic environments, resulting in a complex trade-off. TE-host relationships, as described by population genetics studies, have been shaped not only through conflicts, but also by coevolution processes.

4
Conclusion

Transposable elements appear to be particularly complicated biological entities. Several dozens of TE families show various transposition mechanisms and different dynamic properties, which do not always correspond to their classification. The dynamics of phylogenetically distant families can indeed be more similar than closely related ones, due to fortuitous causes or perhaps because of convergent molecular evolution (Hua-Van et al. 2005). In order to get a better understanding of their evolution, some simple models have been defined. TEs are then characterized by a small number of key-parameters, such as their duplication rate, their deletion rate, and their impact on the host's fitness. Even if these factors appear to be oversimplified, operating in a single model they provide insightful information (Charlesworth et al. 1994; Le Rouzic and Deceliere 2005), raising further biological and evolutionary issues.

TE models generally focus on TE properties. Nevertheless, the host's features may also play a considerable role in the dynamics of their intra-genomic inhabitants. Some species or groups of species do not seem to be prone to invasion by several TE families. For instance, despite the existence of reverse

---

[6] *Exaptations* are features coopted for a current utility following an origin for a different function (or no function at all).

transcriptase-encoding *retrons* in prokaryotes (Inouye et al. 1987), there are no retro-transposons (Class I elements) in bacteria, all insertion sequences being "cut and paste" Class II elements. On the other hand, there are only few Class II elements in yeasts or in primates, even if they are potentially active in these genomes (Izsvak and Ivics 2004). What are the populational, environmental, phylogenetic, genomic, or random factors supposed to explain such differences? The question remains open.

It is now generally accepted that the mode of reproduction has a vital influence on TE biology. The invasion of a selfish DNA sequence is indeed much easier in a sexual population, and the importance of sex in the spread of TEs and in the evolution of TE regulation has been frequently suggested (Zeyl et al. 1996; Bestor 1999, 2003; Arkhipova and Meselson 2000, 2005; Xu and Deng 2002). Moreover, even if the organism reproduces sexually, self-fertilization (in plants for example) can modify the invasion dynamics and induce important TE-content differences between closely related species (Wright and Schoen 1999; Morgan 2001). Finally, ecological differences may also lead to discrepancies between species (Vieira and Biémont 2004), and populational demographic events also probably interfere with TE dynamics (Vieira et al. 1999).

Nevertheless, a few species (generally unicellular eukarya) seem to be totally deprived of TE sequences (*Plasmodium*: Holt et al. 2002; Carlton et al. 2002, *Cryptosporidium*: Abrahamsen et al. 2004; Xu et al. 2004, or microsporidia: Katinka et al. 2001). Their way of life (parasitism), the role of the population structure and migrations between meta-populations, or the interspecific network in which the members of a TE family can evolve, leaping from one species to another through horizontal transfers, are likely to have an important but still misunderstood impact on TE maintenance and long-term evolution.

Acknowledgements We would like to thank D. Lankenau and two anonymous referees for their useful comments. The English text was reviewed by M. Eden.

References

Abrahamsen M, Templeton T, Enomoto S, Abrahante J, Zhu G, Lancto C, Deng M, Liu C, Widmer G, Tzipori S et al. (2004) Complete genome sequence of the apicomplexan, *Cryptosporidium parvum*. Science 304:441–445

Agrawal A, Eastman QM, Schatz DG (1998) Transposition mediated by *RAG1* and *RAG2* and its implications for the evolution of the immune system. Nature 394:744–751

Andrews JD, Gloor GB (1995) A role for the *KP* Leucine Zipper in regulating *P* element transposition in *Drosophila melanogaster*. Genetics 135:81–95

Anxolabéhère D, Benes H, Nouaud D, Périquet G (1987) Evolutionary steps and transposable elements in *Drosophila melanogaster*: the missing *RP* type obtained by genetic transformation. Evolution 4:846–853

Anxolabéhère D, Kidwell MG, Périquet G (1988) Molecular characteristics of diverse populations are consistent with the hypothesis of a recent invasion of *Drosophila melanogaster* by mobile *P* elements. Mol Biol Evol 5:252–269

Aravind L (2000) The BED finger, a novel DNA-binding domain in chromatin-boundary-element-binding proteins and transposases. Trends Biochem Sci 25:421–423

Arkhipova I, Meselson M (2000) Transposable elements in sexual and anciant asexual taxa. Proc Natl Acad Sci USA 97:14473–14477

Arkhipova I, Meselson M (2005) Deleterious transposable elements and the extinction of asexuals. Bioessays 27:76–85

Badge RM, Brookfield JFY (1997) The role of host factors in the population dynamics of selfish transposable elements. J Theor Biol 187:261–271

Bestor T (2003) Cytosine methylation mediates sexual conflict. Trends Genet 19:185–190

Bestor TH (1999) Sex brings transposons and genomes into conflict. Genetica 107:289–295

Biémont C (1994) Dynamic equilibrium between insertion and excision of *P* elements in highly inbred lines from an *M'* strain of *Drosophila melanogaster*. J Mol Evol 39:466–472

Biémont C, Nardon C, Deceliere G, Lepetit D, Loevenbruck C, Vieira C (2003) World-wide distribution of transposable element copy number in natural populations of *Drosophila simulans*. Evolution 57:159–167

Birchler JA, Pal-Bhadra M, Bhadra U (1999) Less from more: cosuppression of transposable elements. Nature Genet 21:148–149

Bird AP (1997) Does DNA methylation control transposition of selfish elements in the germline? Trends Genet 13:469–470

Boissinot S, Chevret P, Furano A (2000) *L1* (*LINE*-1) retrotransposon evolution and amplification in recent human history. Mol Biol Evol 17:915–928

Bonnivard E, Bazin C, Denis B, Higuet D (2000) A scenario for the *hobo* transposable element invasion, deduced from the structure of natural populations of *Drosophila melanogaster* using tandem TPE repeats. Genet Res 75:13–23

Bregliano JC, Kidwell MG (1983) Mobile Genetic Elements. chap. Hybrid Dysgenesis Determinants, p 363. Academic Press, inc.

Brookfield J (1982) Interspersed repetitive DNA sequences are unlikely to be parasitic. J Theor Biol 94:281–299

Brookfield JFY (1995) Mobile genetics elements. chap. Transposable elements as selfish DNA. Oxford University Press, New York, p 131–153

Brookfield J (2003) Mobile DNAs: the poacher turned gamekeeper. Curr Biol 13:R846–R847

Brookfield JFY, Badge RM (1997) Population genetics models of transposable elements. Genetica 52:281–294

Brosius J (2005) Disparity, adaptation, exaptation, bookkeeping, and the contingency at the genome level. Paleobiology 31:1–16

Bucheton A (1990) *I* transposable elements and *I-R* hybrid dysgenesis in *Drosophila*. Trends Genet 6:16–21

Burke WB, Malik HS, Lathe WC, Eickbush TH (1998) Are retrotransposons long-term hitchhikers? Nature 392:141–142

Cambareri EB, Jensen BC, Schabtach E, Selker EU (1989) Repeat-induced G–C to A–T mutations in *Neurospora crassa*. Science 244:1571–1575

Capy P, Gasperi G, Biémont C, Bazin C (2000) Stress and transposable elements: co-evolution or useful parasites? Heredity 85:101–106

Carlton J, Angiuoli S, Suh B, Kooij T, Pertea M, Silva J, Ermolaeva M, Allen J, Selengut J, Koo H et al. (2002) Genome sequence and comparative analysis of the model rodent malaria parasite *Plasmodium yoelii yoelii*. Nature 419:512–519

Charlesworth B (1991) Transposable elements in natural populations with a mixture of selected and neutral insertion sites. Genet Res 57:127–134

Charlesworth B (1996) Background selection and patterns of genetic diversity in *Drosophila melanogaster*. Genet Res 68:131–149

Charlesworth B, Charlesworth D (1983) The population dynamics of transposable elements. Genet Res 42:1–27

Charlesworth B, Langley CH (1986) The evolution of self-regulated transposition of transposable elements. Genetics 112:359–383

Charlesworth B, Sniegowski P, Stephan W (1994) The evolutionary dynamics of repetitive DNA in eukaryotes. Nature 371:215–220

Coen D, Lemaitre B, Delattre M, Quesneville H, Ronsseray S, Simonelig M, Higuet D, Lehmann M, Montchamp C, Nouaud D (1994) *Drosophila P* element: transposition, regulation and evolution. Genetica 93:61–78

Corish P, Black D, Featherston D, Merriam J, Dover G (1996) Natural repressors of *P*-induced hybrid dysgenesis in *Drosophila melanogaster*: a model for repressor evolution. Genet Res 67:109–121

Davis C, Wurdack K (2004) Host-to-parasite gene transfer in flowering plants: phylogenetic evidence from Malpighiales. Science 305:676–678

Deceliere G, Charles S, Biémont C (2005) The dynamics of transposable elements in structured populations. Genetics 169:467–474

Doolittle W, Sapienza C (1980) Selfish genes, the phenotype paradigm and genome evolution. Nature 284:601–603

Doolittle WF, Kirkwood TBL, Dempster MAH (1984) Selfish DNA with self-restraint. Nature 307:501–502

Eanes WF, Wesley C, Hey J, Houle D (1988) The fitness consequences of *P* element insertion in *Drosophila melanogaster*. Genet Res 52:17–26

Eggleston W, Johnson-Schlitz D, Engels W (1988) *P-M* hybrid dysgenesis does not mobilize other transposable element families in *Drosophila melanogaster*. Nature 331:368–370

Engels WR (1989) Mobile DNA. chap. *P* elements in *Drosophila melanogaster*, p 437–484. American Society of Microbiology

Engels W (1997) Invasions of *P* elements. Genetics 145:11–15

Feschotte C, Mouchès C (2000) Evidence that a family of miniature inverted-repeat transposable elements (*MITEs*) from the *Arabidopsis thaliana* genome has arisen from a *pogo*-like DNA transposon. Mol Biol Evol 17:730–737

Ganko E, Bhattacharjee V, Schliekelman P, McDonald J (2003) Evidence for the contribution of LTR retrotransposons to *C. elegans* gene evolution. Mol Biol Evol 20:1925–1931

Ganko E, Fielman K, McDonald J (2001) Evolutionary history of *Cer* elements and their impact on the *C. elegans* genome. Genome Res 11:2066–2074

Gerasimova TI, Mizrokhi LJ, Georgiev GP (1984) Transposition bursts in genetically unstable *Drosophila melanogaster*. Nature 309:714–716

Gould SJ (2002) The structure of evolutionary theory. Belknap Press of Harvard University Press, Cambridge, Mass

Gloor et al. (1993) Type I repressors of *P* element mobility. Genetics 135:81–95

Hickey DA (1982) Selfish DNA: a sexually-transmitted nuclear parasite. Genetics 101:519–531

Hood M, Katawczik M, Giraud T (2005) Repeat-induced point mutation and the population structure of transposable elements in *microbotryum violaceum*. Genetics 170:1081–1089

Holt R, Subramanian G, Halpern A, Sutton G, Charlab R, Nusskern D, Wincker P, Clark A, Ribeiro J, Wides R et al. (2002) The genome sequence of the malaria mosquito *Anopheles gambiae*. Science 298:129–149

Houle D, Nuzhdin SV (2004) Mutation accumulation and the effect of *copia* insertions in *Drosophila melanogaster*. Genet Res 83:7–18

Hua-Van A, Le Rouzic A, Maisonhaute C, Capy P (2005) Abundance, distribution and dynamics of retrotransposable elements and transposons: similarities and differences. Cytogenet Genome Res 110:426–440

Inouye S, Furuichi T, Dhundale A, Inouye M (1987) Molecular biology of RNA: new perspectives. chap. Stable branched RNA covalently linked to the 5′ end of a single-stranded DNA of *Myxobacteria*, p 271–284. Academic Press, San Diego/London

Izsvak Z, Ivics Z (2004) Sleeping beauty transposition: biology and applications for molecular therapy. Mol Ther 9:147–156

Jackson M, Black D, Dover G (1988) Amplification of *KP* elements associated with the repression of hybrid dysgenesis in *Drosophila melanogaster*. Genetics 120:1003–1013

Jordan I, Rogozin I, Glazko G, Koonin E (2003) Origin of a substantial fraction of human regulatory sequences from transposable elements. Trends Genet 19:68–72

Kaplan N, Darden T, Langley CH (1985) Evolution and extinction of transposable elements in mendelian populations. Genetics 109:459–480

Katinka M, Duprat S, Cornillot E, Metenier G, Thomarat F, Prensier G, Barbe V, Peyretaillade E, Brottier P, Wincker P et al. (2001) Genome sequence and gene compaction of the eukaryote parasite *Encephalitozoon cuniculi*. Nature 414:450–453

Kazazian H (2000) *L1* retrotransposons shape the mammalian genome. Science 289:1152–1153

Kidwell M (1983) Evolution of hybrid dysgenesis determinants in *Drosophila melanogaster*. Proc Natl Acad Sci USA 80:1655–1659

Kidwell MG (1992) Horizontal transfer. Curr Opin Genet Dev 2:868–873

Kidwell MG, Lisch DR (2001) Transposable elements, parasitic DNA, and genome evolution. Evolution 55:1–24

Kurland C, Canback B, Berg O (2003) Horizontal gene transfer: a critical view. Proc Natl Acad Sci USA 100:9658–9662

Labrador M, Corces VG (1997) Transposable element-host interactions: regulation of insertion and excision. Annu Rev Genet 31:381–404

Lampe D, Witherspoon D, Soto-Adames F, Robertson H (2003) Recent horizontal transfer of mellifera subfamily *mariner* transposons into insect lineages representing four different orders shows that selection acts only during horizontal transfer. Mol Biol Evol 20:554–562

Langley CH, Montgomery E, Hudson R, Kaplan N, Charlesworth B (1988) On the role of unequal exchange in the containment of transposable element copy number. Genet Res 52:223–235

Lankenau S, Corces VG, Lankenau DH (1994) The *Drosophila micropia* retrotransposon encodes a testis-specific antisense RNA complementary to reverse transcriptase. Mol Cell Biol 14:1764–1775

Lankenau DH, Gloor GB (1998) In vivo gap repair in *Drosophila*: a one-way street with many destinations. BioEssays 20:317–327

Le Rouzic A, Capy P (2005) The first steps of transposable elements invasion: parasitic strategy *vs.* genetic drift. Genetics 169:1033–1043

Le Rouzic A, Deceliere G (2005) The models of population genetics of transposable elements. Genet Res 85:171–181

Leaver M (2001) A family of *Tc1*-like transposons from the genomes of fishes and frogs: evidence for horizontal transmission. Gene 271:203–214

Lemaitre B, Ronsseray S, Coen D (1993) Maternal repression of the *P* element promoter in the germline of *Drosophila melanogaster*: a model for the *P* cytotype. Genetics 135:149–160

Lippmann Z, May B, Yordan C, Singer T, Martienssen R (2003) Distinct mechanisms determine transposon inheritance and methylation via small interfering DNA and histone modification. PLoS Biol 1:420–428

Mackay TFC, Lyman R, Jackson M (1992) Effects of *P* element insertions on quantitative traits in *Drosophila melanogaster*. Genetics 130:315–332

Mackay TFC (1985) Transposable element-induced response to artificial selection in *Drosophila melanogaster*. Genetics 111:351–374

Marin L, Lehmann M, Nouaud D, Izaabel H, Anxolabehere D, Ronsseray S (2000) *P*-element repression in *Drosophila melanogaster* by a naturally occurring defective telomeric *P* copy. Genetics 155:1841–1854

Martienssen R (1998) Transposons, DNA methylation and gene control. Trends Genet 14:263–264

Maside X, Assimacopoulos S, Charlesworth B (2000) Rates of movement of transposable elements on the second chromosome of *Drosophila melanogaster*. Genet Res 75:275–284

Miller W, McDonald J, Nouaud D, Anxolabéhère D (1999) Molecular domestication – more than a sporadic episode in evolution. Genetica 107:197–207

Montchamp-Moreau C (1990) Dynamics of *P-M* hybrid dygenesis in *P*-transformed lines of *Drosophila simulans*. Evolution 44:194–203

Morgan MT (2001) Transposable element number in mixed mating populations. Genet Res 77:261–275

Nuzhdin SV (1999) Sure facts, speculations, and open questions about evolution of transposable elements. Genetica 107:129–137

Nuzhdin SV, Mackay TFC (1995) The genomic rate of transposable element movement in *Drosophila melanogaster*. Mol Biol Evol 12:180–181

Ohta T (1986) Population genetics of an expanding family of mobile genetic elements. Genetics 113:145–159

Orgel LE, Crick FHC (1980) Selfish DNA: the ultimate parasite. Nature 284:604–607

Pardue ML, DeBaryshe P (2003) Retrotransposons provide an evolutionnary robust non-telomerase mechanism to maintain telomeres. Annu Rev Genet 37:485–511

Petrov DA (2001) Evolution of genome size: new approches to an old problem. Trends Genet 17:23–28

Petrov D, Hartl D (1998) High rate of DNA loss in the *Drosophila melanogaster* and *Drosophila viridis* species group. Mol Biol Evol 15:293–302

Petrov D, Schutzman J, Hartl D, Lozovskaya E (1995) Diverse transposable elements are mobilized in hybrid dysgenesis in *Drosophila virilis*. Proc Natl Acad Sci USA 92:8050–8054

Quesneville H, Anxolabéhère D (1998) Dynamics of transposable elements in metapopulations: a model of *P* elements invasion in *Drosophila*. Theor Pop Biol 54:175–193

Quesneville H, Anxolabéhère D (2001) Genetic algorithm-based model of evolutionnary dynamics of class II transposable elements. J Theor Biol 213:21–30

Quesneville H, Nouaud D, Anxolabéhère D (2003) Detection of new transposable element families in *Drosophila melanogaster* and *Anopheles gambiae* genomes. J Mol Evol 57 Suppl 1:S50–S59

Rio DC (1991) Regulation of *Drosophila P* element transposition. Trends Genet 7:282–287

Roussigne M, Kossida S, Lavigne A, Clouaire T, Ecochard V, Glories A, Amalric F, Girard J (2003) The *THAP* domain: a novel protein motif with similarity to the DNA-binding domain of *P* element transposase. Trends Biochem Sci 28:66–69

San Miguel P, Tikhonov A, Jin YK, Motchoulskaia N, Zakharov D, Melake-Berhan A, Springer PS, Edwards KJ, Lee M, Avramova Z et al. (1996) Nested retrotransposons in the intergenic regions of the maize genome. Science 274:765–768

Sanchez-Gracia A, Maside X, Charlesworth B (2005) High rate of horizontal transfer of transposable elements in *Drosophila*. Trends Genet 21:200–203

Silva J, Loreto E, Clark J (2004) Factors that affect the horizontal transfer of transposable elements. Curr Issues Mol Biol 6:57–71

Simmons M, Raymond J, Rasmusson K, Miller L, McLarnon C, Zunt J (1990) Repression of *P* element-mediated hybrid dysgenesis in *Drosophila melanogaster*. Genetics 124:663–676

Snyder M, Doolittle W (1988) *P* elements in *Drosophila*: selection at many levels. Trends Genet 4:147–149

Suh D, Choi E, Yamazaki T, Harada K (1995) Studies on the transposition rates of mobile genetic elements in a natural population of *Drosophila melanogaster*. Mol Biol Evol 12:748–758

Tsitrone A, Charles S, Biémont C (1999) Dynamics of transposable elements under the selection model. Genet Res 74:159–164

Vieira C, Biémont C (2004) Transposable element dynamics in two sibling species: *Drosophila melanogaster* and *Drosophila simulans*. Genetica 120:115–123

Vieira C, Lepetit D, Dumont S, Biemont C (1999) Wake up of transposable elements following *Drosophila simulans* worldwide colonization. Mol Biol Evol 16:1251–1255

Vieira J, Vieira C, Hartl D, Lozovskaya E (1998) Factors contributing to the hybrid dysgenesis syndrome in *Drosophila virilis*. Genet Res 71:109–117

Vitte C, Panaud O (2003) Formation of solo-LTRs through unequal homologous recombination counterbalances amplifications of LTR retrotransposons in rice *Oryza sativa* L. Mol Biol Evol 20:528–540

Weiner A (2002) *SINEs* and *LINEs*: the art of biting the hand that feeds you. Curr Opin Cell Biol 14:343–350

Witherspoon D (1999) Selective constraints on *P*-element evolution. Mol Biol Evol 16:472–478

Woodruff R, Thompson J Jr, Gu S (2004) Premeiotic clusters of mutation and the cost of natural selection. J Hered 95:277–283

Wright SI, Schoen DJ (1999) Transposon dynamics and the breeding system. Genetica 107:139–148

Xu P, Widmer G, Wang Y, Ozaki L, Alves J, Serrano M, Puiu D, Manque P, Akiyoshi D, Mackey A et al. (2004) The genome of *Cryptosporidium hominis*. Nature 431:1107–1112

Xu T, Deng K (2002) Sex and retrotransposons: a new approach to the problem. J Theor Biol 218:259–260

Yoder JA, Walsh CP, Bestor TH (1997) Cytosine methylation and the ecology of intragenomic parasites. Trends Genet 13:335–340

Zeyl C, Bell G, Green DM (1996) Sex and spread of retrotransposon *ty3* in experimental populations of *Saccharomyces cerevisiae*. Genetics 143:1567–1577

Genome Dyn Stab (4)
D.-H. Lankenau, J.-N. Volff: Transposons and the Dynamic Genome
DOI 10.1007/7050_2009_044/Published online: 25 March 2009
© Springer-Verlag Berlin Heidelberg 2009

Infra- and Transspecific Clues to Understanding the Dynamics of Transposable Elements

Cristina Vieira (✉) · Marie Fablet · Emmanuelle Lerat

Laboratoire de Biométrie et Biologie Evolutive, Université de Lyon; Université Lyon 1; CNRS; UMR 5558, 69622 Villeurbanne, France
vieira@biomserv.univ-lyon1.fr

Abstract All genomes contain, to a greater or lesser extent, sequences that do not seem to be beneficial. The most preeminent group consists of transposable elements (TEs). These repeated DNA sequences have a significant influence on genome dynamics and evolution. One of the main challenges facing modern molecular evolution is to understand and measure their impact on evolution. The aim of this paper is to establish the relevance and contribution of population studies, as well as the species comparative approaches, to understanding the dynamics of TEs. Most of the examples cited concern the species *Drosophila melanogaster*, since this is one of the genetic key-model organisms, for which an enormous amount of data has been collected over a period of 100 years of genetic research, and which represents a genus for which the genomes of 12 species have been sequenced.

Abbreviations

TE	Transposable element
LINE	Long interspersed nuclear element
LTR	Long terminal repeat
UTR	Untranslated region
RNAi	RNA interference
siRNA	Small interfering RNA
rasiRNA	Repeat-associated small interfering RNA

1
Introduction

Historically, the conventional view of genome evolution has associated organism complexity with the number of protein coding genes. However, the recent complete sequencing of the human genome has shown how similar it is to that of *Drosophila*, since the human genome only has twice its gene number (Lander et al. 2001), and this has highlighted the relevance of non-protein coding gene pathways in controlling the differentiation and diversity of organisms (Taft and Mattick 2003). Another important finding arising from this sequencing program is that only 2% (Goodstadt and Ponting 2006; Human

Genome Sequencing Consortium 2004) of the human genome actually codes for proteins, the function of the rest being still unknown. Large-scale studies have shown that some non-coding regions are very well conserved across species, which suggests that they must in fact have some "function" (Bejerano et al. 2004). In the human genome, 42% of these "non protein-coding gene" regions are constituted by transposable elements (TEs) (Human Genome Sequencing Consortium 2004). This inevitably raises the question of the role of TEs in genome evolution and makes them indwelling components of genomes (see Walisko et al., in this volume).

For a long time scientists thought that the genome was a stable entity, and it was only in the 1950s, thanks to the work of McClintock, that doubts began to trouble this supposedly calm landscape (McClintock 1984, Fedoroff and Botstein 1992). The genome fluidity we now consider as obvious, was difficult to accept at the time. The first TEs were discovered in maize (McClintock 1950), but most of the subsequent early work was done on bacteria (Shapiro 1969, Saedler and Starlinger 1992), since they were a lot easier to study at the molecular level. The first TEs to be described were DNA transposons, i.e., elements that transpose via a DNA intermediate. Studies in *Drosophila*, *Caenorhabditis elegans*, and other eukaryotes, subsequently identified RNA elements, i.e., elements that transpose using an RNA intermediate. Herein, we make no claim to discuss the precise classification of TEs, both because several different systems are possible, and also because new elements are reported every day (Kapitonov and Jurka 2008; Wicker et al. 2007). We will therefore adopt the former classification proposed by Finnegan (1989), which distinguishes two major classes of TEs, based on their transposition cycle intermediates.

TEs are DNA sequences that encode the enzymes necessary for their transposition, i.e., to allow them to move between non-homologous regions in the genomes or to copy themselves to other positions. In some cases, TEs known as non-autonomous sequences do not produce their own enzymes, but are able to use those from functional copies or even from other TE families. The amount of TEs and its impacts on genome stability vary widely among organisms. For instance, retrotransposons constitute almost one half of the human genome, but they are responsible for only 0.2% of spontaneous mutations (Kazazian 1998), while in *Drosophila*, for which the TE contribution is much reduced in terms of genome occupancy, TEs are proposed to be the source of more than 50% of spontaneous mutations with notable effects (Eickbush and Furano 2002). Transposition rates may thus be higher than spontaneous mutation rates, as in *Drosophila*, in which these rates are estimated to be 10^{-3}–10^{-4} (Vieira and Biémont 1997; Suh et al. 1995; Nuzhdin and Mackay 1995) and 10^{-8} (Crow and Simmons 1983), respectively.[1] The evolution of new

---

[1] These are global values for the transposition rates, independently of mutation causes such as double-stranded breaks, as suggested by W.D. Heyer in the third volume of this collection.

insertions in a genome should be considered at two time scales. The short-term effects will depend on the insertion site; if the insertion disrupts a gene and consequently affects the fitness of the organism, we can expect it to be eliminated by natural selection, whereas if the insertion is in a non-coding region, we may expect it to be maintained if it has no impact on host fitness.[2] Long-term effects will only involve insertions that are associated with very weak deleterious effects (Langley et al. 1988), since these are the only ones not promptly eliminated. This makes it possible to identify fixed insertions in populations, which may not necessarily be adaptive, but can simply be the consequence of genetic drift and bottlenecks (Cordaux et al. 2006a; Cordaux et al. 2006b). Furthermore, insertions of TEs may modify regulatory pathways and the expression patterns of genes when they insert in their vicinity (Peaston et al. 2004), and may also be subject to strong selection, leading to an increase in their frequency in populations and enhanced host fitness (Aminetzach et al. 2005). The occurrence of molecular domestication events is now frequently reported, and seems to happen in many different organisms (Feschotte and Pritham 2007; Kapitonov and Jurka 2005; Miller et al. 1997), implying that TEs play a key role in genome evolution.[3]

We describe here the way population-based studies and species comparative analyses have contributed to the current understanding of TE dynamics and evolution, focusing on different levels of study of TEs, from the copy number, to sequence variation, and the epigenetic regulation of activity.

2
Lessons from the Past

2.1
The Heritage of Hybrid Dysgenesis Studies in *Drosophila* Populations

After their discovery by Barbara McClintock in the 1950s, TE study went through a new birth in the late 1970s when drosophilists related aberrant traits in some crosses of *Drosophila melanogaster* strains. Among these aberrant traits were recombination in males (Hiraizumi et al. 1973) – which is not expected to occur in *D. melanogaster* –, high rate of mutation (Thompson and Woodruff 1980), sterility, chromosomal aberrations (Kidwell et al.

---

[2] Here, we only refer to "regular", punctual transposition events, as opposed to the massive bursts of transposition observed in *Drosophila* in the case of what is called hybrid dysgenesis (Kidwell et al. 1977), which will be developed in Sect. 2.1. This phenomenon is observed when crossing individuals originating from strains differing in their TE content, and results in a high rate of mutation, chromosome rearrangements, and sterility in the offspring, due to an extremely elevated rate of transposition. In this case, even if the insertion sites are not located in coding regions, the effects on the offspring fitness are considerable.

[3] For an extensive review on domestication of TEs, refer to Dettai and Volff, in this volume, and Volff 2006.

1977). These aberrations were found non reciprocally in F1 hybrids, and some of the traits were even not found in non hybrids. This led Margaret Kidwell and colleagues (1977) to use the term "hybrid dysgenesis" to qualify such a phenomenon. *D. melanogaster* strains could be classified into two types, called P and M, according to the paternal or maternal contribution in the production of hybrid dysgenesis. It appeared that strains collected from natural populations at that time were typically of the P type and those having a long laboratory history were of the M type (Kidwell et al. 1977). At the same time, Picard (1976) reported another system of hybrid dysgenesis, distinguishing inducer (I), reactive (R), or neutral (N) strains. All strains collected from the wild were classified as I strains. Geneticists at that time proposed that all of these aberrant traits could be related, and caused by chromosomal factors, but their identification proved to be hard due to the difficulty in localizing the causal factor(s) to a single chromosome (Kidwell et al. 1977). It was subsequently considered that hybrid dysgenesis in the P-M system resulted from the interaction of a chromosomal component ("P factor") and an extrachromosomal property ("M cytotype") (Engels and Preston 1980). This P factor actually corresponds to the now well-studied P transposon, and the I-R hybrid dysgenesis proved to be due to another transposable element, the I non-LTR retrotransposon.

Studies of the hybrid dysgenesis phenomenon proved that the invasion of a genome by TEs was possible and could happen in a relatively short time. This motivated the approach of TEs by modeling, so that in the early 1980s, several authors proposed theoretical models intended to explain the dynamics of TEs (Le Rouzic and Decelière 2005 for a review). These relatively simple models could be used to test neutrality or selection of the deleterious effects of TE insertion, or the effects of recombination induced by TEs. The value of a model depends on being able to test it. In this respect, *Drosophila* is a very suitable model organism for such tests. In fact, *Drosophila* is unmatched in two characteristics: (1) the giant polytene chromosomes and (2) balancer chromosomes. One most prominent experimental tool distinguishing *Drosophila* from other model systems is the advantage to be able to carry out in situ hybridizations on polytene chromosomes (Gall and Pardue 1969; Pardue and Gall 1969), and map the sites as well as determine the copy number (Biémont et al. 2004, Fig. 1) by means of the classical and still very useful chromosome maps of Bridges (Bridges 1935).

The first studies were done on laboratory populations of *Drosophila* (Langley et al. 1988), and then several studies were performed on natural populations (Biémont 1994; Biémont et al. 1994; Hoogland and Biémont 1996). As has been demonstrated in several reviews, no general model can be applied to all TEs and all populations, since they both are rarely at equilibrium (Biémont et al. 1997). Further, the precise biochemical details of transposition of a TE family in general and each individual TE specifically, embedded in its particular chromatin environment, is different in each specific circumstance. Copy number

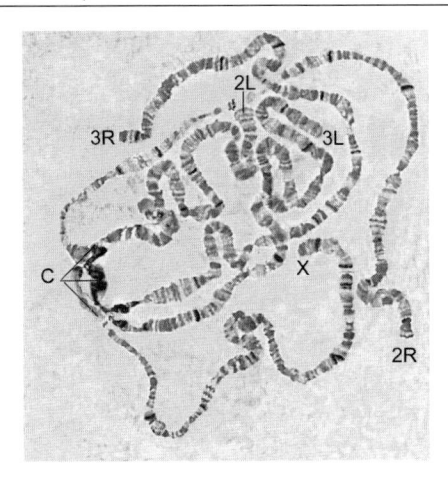

Fig. 1 In situ hybridization on *Drosophila* salivary gland polytene chromosomes. The hybridization was performed with a biotinylated DNA probe (reviewed by Biémont et al. 2004). The probe, with sequence homology to the 412 LTR retrotransposon, is detected as multiple black bands on the chromosome preparation. The position of each TE can be precisely identified and linked to the maps of the complete *Drosophila* genome. C: chromocenter, 2L and 2R are the left and right arms of the chromosome 2, 3L and 3R left and right arms of the chromosome 3

data obtained by in situ hybridization in *D. melanogaster* were not easy to extrapolate to other species of *Drosophila*, even to closely related species such as *D. simulans* (Vieira and Biémont 1996, 2004; Vieira et al. 2000). Comprehensive analyses of numerous individuals and populations soon became impossible to manage practically. In addition, one of the main problems with the in situ approach is the approximate nature of the localizations. It is quite difficult to distinguish between neighboring sites, and also to be sure of the sequence similarity between the probes and the highlighted spots. This made it impossible to identify all potentially fixed sites, leading to the conclusion that insertion polymorphism levels in *Drosophila* were high. Using the insertion sites detected in the sequenced genome and searching for them in individuals in a natural population, led to the identification of numerous fixed insertions. The evolutionary significance of these insertions is still under investigation, and we need to be able to distinguish between genetic drift and adaptive selection (Aminetzach et al. 2005; Lipatov et al. 2005; Macpherson et al. 2008; McCollum et al. 2002; Dettai and Volff, in this volume).

2.2
The Sibling Species *D. melanogaster* and *D. simulans*

As previously mentioned herein, the number of TE copies varies extensively when considering different model genomes, such as *D. melanogaster*, *Homo sapiens*, or *Zea mais*. Nevertheless, one might have expected that closely re-lated species had the same copy number of TEs – at least of the same order of magnitude, if we assume that these species have been submitted to similar evolutionary processes. However, this assumption turned out to be incorrect, as was revealed by the analysis of two sibling species of the genus *Drosophila*, *D. melanogaster* (the model species of metazoan genetics) and *D. simulans*. It has been shown that the copy number of most TEs (obtained by in situ hybridization) is smaller in *D. simulans* than in *D. melanogaster*. But more surprisingly, there is a huge difference in copy numbers between natural populations of *D. simulans* with regard to several TE families. In fact, most populations have very low copy numbers, but there are a few exceptions, in which the copy number is very high, sometimes even higher than the aver-age value found in *D. melanogaster*. This has led us to hypothesize that the genome of *D. simulans* is beginning to be invaded by TEs, and that this in-vasion could be associated with the current worldwide colonization of the *D. simulans* species (Biémont et al. 2003; Vieira and Biémont 2004; Vieira et al. 1999). However, the alternative hypothesis of an ancient invasion followed by a progressive loss of TEs cannot be ruled out. Recent data obtained for a LINE element and an LTR retrotransposon seem to support the latter hy-pothesis (Fablet et al. 2006; Rebollo et al. 2008). The main challenges facing us are understanding why some populations are sporadically invaded by a spe-cific family of TEs, identifying the genetic and/or environmental factors that have allowed this invasion, and finding out how TEs are eliminated.

2.3
In the Genome Sequencing Era

The recent explosion of sequencing projects, making ever more genomes available, is an important step forward in determining the TE loads of dif-ferent species. The analysis of the genomes of 12 *Drosophila* species and the genomes from other insects, such as *Anopheles gambiae* (Holt et al. 2002), *Aedes aegypti* (Nene et al. 2007), *Pediculus humanus, Bombyx mori* (Xia et al. 2004), *Tribolium castaneaum* (Tribolium Genome Sequencing Consortium 2008), *Nasonia vitripennis*, or *Apis mellifera* (Honeybee Genome Sequenc-ing Consortium 2006), has shown that there are significant differences in the amount, type and degree of conservation of TEs between different species. The genome of *Apis* differs from previously sequenced insect genomes in that it presents very small amounts of TEs, with especially very few retrotrans-posons (Honeybee Genome Sequencing Consortium 2006). Most of the TEs in *Apis* are from the *mariner* family, a DNA transposon, whereas other types of transposons and retrotransposons are present, but only as highly degraded copies, indicating that they are no longer active. In contrast, the silkworm has a very large genome, of which TEs account for 21.1%, and it is probably TEs that are responsible for the increase in genome size in this species (Xia et al. 2004). Most of these TEs mainly belong to one family of LTR retrotransposons. The genome of *Anopheles* is also larger than that of *D. melanogaster*, and in this species TEs account for about 16% of the euchromatin with a good representation of different families from DNA transposons and retrotransposons (Holt et al. 2002).

The analyses of other eukaryotic genomes have demonstrated very different patterns of TE dynamics. For example, traces of waves of amplification/loss over evolutionary time have been identified in the human genome, where the *L1* elements (non-LTR retrotransposons) are the ones that have invaded the genome the most recently. Furthermore, several observations (Boissinot and Furano 2005; Mathews et al. 2003) indicate that competition exists between different LINE 1 families, which could explain why only one subfamily is now active in the human genome. The amplification of TEs has been reported in many plants, and is responsible for the increase in the genome size of maize (SanMiguel et al. 1998).

However, despite the large amount of data that can be obtained from analyses of sequenced genomes, specific population studies addressing specifically particular TEs and their molecular characteristics of transposition are still the only way to approach some particular issues. For instance, Herrera et al. (2006) recently used PCR analysis to investigate the presence/absence of four insertions of the *HERV-K* element in nine human populations from various geographical origins. The authors observed a general discrepancy from Hardy–Weinberg equilibrium expectations, which they propose might be due to the fact that the heterozygotes are subjected to pandemic negative selection. These observations are of great interest in attempting to understand the selective pressures that drive TE dynamics within a germline genome.

3
Towards an Understanding of TE Regulation. From Sequence to Epigenetics

3.1
Sequence Variability

The other way to analyze TE dynamics is to compare their intrinsic sequence characteristics. This allows us to address questions regarding the phylogeny and evolution of individual TEs. These types of analyses are based on the assumption that the evolution rates of TEs are the same as those of conventional genes, which of course is not always true (Malik et al. 1999; Malik and Eickbush 1999, 2001). One of the most striking conclusions of the phylogeny studies from sequenced genomes was that TEs evolve independently of their host genomes (Lerat et al. 2002a,b). Phylogenetic studies of the same TEs in different species have demonstrated that horizontal transfers can occur (Clark et al. 1994). These studies were mainly based on coding regions, and there have been very few population studies (Silva et al. 2004).

On the other hand, several population studies have been carried out on non-translated regions, such as the long terminal repeat (LTR) and untranslated region (UTR). LTR-retrotransposons are the most common TEs in eukaryotic organisms, especially in *Drosophila*, in which more than 2/3 of the TEs belong to this class (Hoskins et al. 2007). These TEs transpose via an RNA intermediate, and are dependent on reverse transcriptase activity. The nature of the replication process of these retrovirus-like TEs favors the appearance of duplications that may behave as enhancers during the evolutionary process (McDonald et al. 1997). Such enhancers may act in *cis* on the element itself, or may influence other genes in the vicinity of their point of insertion (Peaston et al. 2004). In recent years several studies have attempted to characterize the regulatory regions of some elements. In *Drosophila*, TEs such as *copia*, *blood*, *412*, *gypsy*, *tirant*, *micropia*, *Ulysses*, and *ZAM* have been extensively studied, and in some cases natural population lines were used. For example, the *copia* element has repetitive motifs in its 5' LTR and 5' UTR[4] regions in *D. melanogaster* (Matyunina et al. 1996), and this region acts as an enhancer in regulating the expression of the element (Wilson et al. 1998). An analysis of several natural populations of *D. melanogaster* and *D. simulans* has revealed the existence of three types of sequence variants in the region spanning from the 5' LTR to the 5' UTR, depending on whether specific portions of the sequence that encompass repeated motifs have been deleted. Two of these *copia* variants are present in both species, but the third occurs only in *D. simulans*. The existence of regulatory region variants has also been reported later for other TEs, such as *blood* (Costas et al. 2001), *412* (Mugnier et al. 2005), and *tirant* (Fablet et al. 2006). In some cases, all the variants seem to be active, but the activity of others such as *copia* and *tirant*, seems to be restricted to a few populations in *D. simulans*.

D. melanogaster populations tend to have rather comparable TE contents (Vieira et al. 1999), but they display considerable variability in their TE regulatory region sequences. For instance, the 5' UTR region of the *tirant* element displays a high variability with regard to the number of tandem repeats of a 102-bp motif among *D. melanogaster* populations worldwide. In addition, investigation of the sequence variability of the *tirant* regulatory region in the four closely related species of the *D. melanogaster* subgroup (*D. melanogaster*, *D. simulans*, *D. sechellia*, and *D. mauritiana*) has revealed different pathways for *tirant* dynamics and evolution, mainly depending on the structure of the populations and on the history of the species (Fablet et al. 2007b). A classi-

---

[4] The typical structure of LTR retrotransposons is made of the genes necessary for their mobilization, bordered by two LTRs, 5' and 3'. Just downstream of the 5' LTR, and 3' to the tRNA primer binding site, a transcribed-untranslated region (UTR) is found. Regulatory sequences are concentrated in LTR and this UTR region (Fablet et al. 2007a).

cal example is the *micropia* retrotransposons of *D. melanogaster* and *D. hydei* (Lankenau et al. 1988; Huijser et al. 1988). While the LTRs of *micropia* within the same species are highly conserved between individual *micropia* elements, the LTRs of *micropia* elements from the other species are completely different except a tandem of putative ecdysteroid receptor binding sites (Lankenau et al. 1990).

Once genomes began to be sequenced, global studies of the amount and structure of the TEs and their dynamics within a given family were possible. The first eukaryotic genome to be sequenced (in 1996) was the yeast *Saccharomyces cerevisiae*, and this was the beginning of global genomic analyses of TEs (Bassett et al. 1996). In particular, Jordan and McDonald (Jordan and McDonald 1998, 1999) analyzed the different families of the *Ty* LTR retrotransposons, and showed that they had recently transposed, and were subject to severe selective pressure. From the analysis of the sequenced genome of *D. melanogaster* (Bowen and McDonald 2001; Lerat et al. 2003), we have learned that the euchromatic insertions of TEs are very recent in this species, and that there is no continuous range of divergence among copies in each class. Indeed, most of the TEs are very similar in sequence to the canonical ones, whereas the other few TEs display high divergence. On the basis of this observation, the hypothesis currently accepted is that in *D. melanogaster*, after intensive TE activity, any degenerate elements are eliminated by a high turnover mechanism. There are of course exceptions, since some very old elements can also be found such as *helena* (Rebollo et al. 2008) or *micropia* (Lankenau et al. 1988). The question as to whether this high turnover mechanism is shared by other *Drosophila* species should soon be resolved, since 11 other *Drosophila* genomes have been recently sequenced (Drosophila 12 Genomes Consortium 2007). Indeed, the analysis of the *D. simulans* sequenced genome has shown, surprisingly, that contrarily to the in situ hybridization data obtained before, this species contains a high number of deleted and rearranged TEs, with few complete sequences for a few families of TEs, which were probably horizontally transmitted into the *D. melanogaster* genome. Moreover, no high turnover is detected in *D. simulans* (Lerat et al., unpublished data).

The initial sequencing programs concentrated on the euchromatic portion of the genomes, since this is the region where active genes have usually been identified. Furthermore, the small amount of highly repeated sequences in these regions makes them much easier to sequence and annotate. Recently, *Drosophila* heterochromatin was sequenced, and it was found that 77% of it is composed of TEs, with few complete elements (Hoskins et al. 2007). Analysis of the LTR proviral-like elements revealed the presence of numerous full-length sequences, which indicates invasion of heterochromatin to be a recent phenomenon (Mugnier et al. 2008). The dynamics of heterochromatin expression is still unknown, but we can envisage several types of gene regulation that could take into account the high proportion of repeated sequences. In addition, it has also been recently demonstrated that some retrotransposons harbored chromodomains in their integrase sequences, which triggers them to heterochromatic regions. These copies then themselves become heterochromatic targets for new subsequent retrotransposon insertions, in an autocatalytic-like fashion (Gao et al. 2008).

The significant efforts devoted to genome sequencing have increased the relevance of comparative genomic approaches, and made it easier to perform gene prediction and annotation. However, from a population genomist's point of view, the sequencing of different species is definitely not enough. The case of *D. melanogaster* is particularly striking, as the sequenced strain, an M strain already used by Morgan in the first genetic studies on this species (Morgan et al. 1915), corresponds to an ancient laboratory line, extinct in the wild. Its genome is clearly different from those found in individuals from natural populations. For example, this genome does not contain any P element insertion, as the strain was collected before the invasion of the genome by this element, whereas all recent natural populations do contain this TE (Engels 1997). We would expect that, in the near future, sequencing projects will consider several individuals of the same species taken from different populations. This approach has already begun with the sequencing of seven lines of *D. simulans* (http://www.dpgp.org/). The first analyses have made it possible to study polymorphism and divergence in this species using a whole genome approach (Begun et al. 2007).

3.2
TE Dynamics at the Epigenetic Level

Epigenetics is an "old" field that is currently experiencing a new flowering.[5] The number of phenomena that may be considered as epigenetic is constantly on the increase. We now realize that all the different phenomena independently reported some years ago, such as chromatin compaction via DNA methylation, or RNA interference, are actually closely linked epigenetic mechanisms. Considerable evidence has been amassed showing that the regulation of TEs (see Fig. 2) is mediated by epigenetic factors, and that this regulation can impact on the regulation of the host gene (Brennecke et al. 2007; Conley et al. 2008; Weil and Martienssen 2008 for a review). Several examples have also shown how TE regulation can depend on environmental conditions. For example, the transposition of P elements in *Drosophila* is temperature sensitive, and indirectly, copy number variation may be associated

---

[5] For an historical perspective on the evolution of the concept of epigenetics, see Jablonka and Lamb 2002. The term is attributed to Conrad Waddington, who used it in the 1940s to refer to "the causal interactions between genes and their products which bring the phenotype into being". The term was then little used in the following decades and reappeared recently, with a slightly different meaning. It now refers to reversible, heritable modifications of genetic expression that are not due to changes in the DNA sequence.

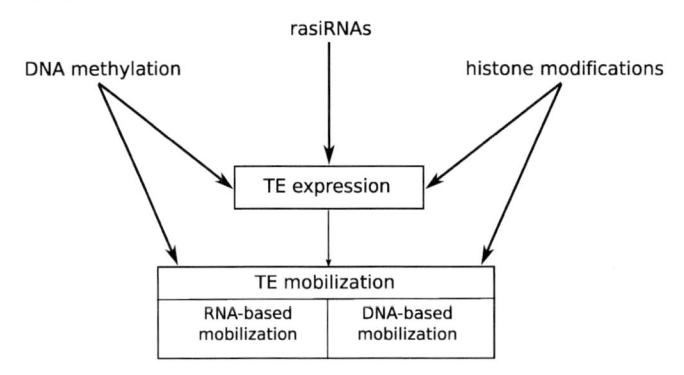

Fig. 2 Schematic view of the different aspects of epigenetic regulation of TEs. Although mechanisms and effects on genome dynamics differ for RNA- and DNA-based TEs, in both cases, mobilization is dependent on TE expression. Several epigenetic pathways may interfere with this expression, such as methylation of cytosines in the DNA molecule, post-translational changes in the histone tails, and RNA interference. These different pathways are interdependent, and a high number of interactions has been described. These epigenetic modifiers have direct effects on gene expression and are responsible for phenotypic changes. In most of the cases, epigenetic modifications are sensitive to the environment, so that genotype X environment interactions can be observed

with environmental factors (Kalendar et al. 2000; Kidwell 1977; Vieira et al. 1998). We are thus tempted to envisage links between epigenetic regulation, TEs and environment. The investigation of natural populations may provide a useful way to address these questions. We will describe here very briefly the three major epigenetic mechanisms known to be implicated in the epigenetic control of TE dynamics.[6]

3.2.1
DNA Methylation

DNA methylation is considered to be the first level of gene regulation. It is an epigenetic mark that is stable and lasting, and it is transmitted throughout development (Reik 2007). In human, de novo methylation is mediated by Dnmt3 (Dnmt3a and Dnmt3b) and methylation is then maintained by Dnmt1. A third family has been described, Dnmt2, which methylates asymmetric cytosines and also presents tRNA methyltransferase activity (Jeltsch et al. 2006). The consequences of cytosine methylation depend on the organism concerned and different types of enzymes are involved depending on the species (see Table 1). For example, in mammals, DNA methylation drives gene imprinting during development, leading to the differential expression of the genes (Hellman and Chess 2007). In plants, DNA methylation

---

[6] An extensive network of researchers working on the epigenome has been established (the Epigenome Network of Excellence). See http://www.epigenome-noe.net/ for abundant data on the epigenetic control.

Table 1 Main enzymes involved in DNA methylation in different groups of organisms

	Enzyme	5meCy rate	Methylation type	Reference
Mammals	DNA methyltransferase family (Dnmt1, 2 and 3)	2–5%	Symmetric	Pradhan and Esteve 2003
Plants	Domains rearranged methyltransferase 2 (Drm2) Methyltransferase 1 (Met1) Dnmt2 Chromomethylase (Cmt3)	19%	Symmetric and asymmetric	Chan, Henderson and Jacobsen 2005
Honeybee	Dnmt family	< 1%	Symmetric	Wang et al. 2006
Drosophila	Dnmt2	< 1%	Asymmetric	Marhold et al. 2004
Zebrafish	Dnmt family	1%	Symmetric	Mhanni and McGowan 2004
Nematode	Not identified	ND		Ponger and Li 2005
Yeast	Not identified	ND		Ponger and Li 2005

is mainly associated with TE silencing (Slotkin and Martienssen 2007). Until recently, *Drosophila* was considered to be a non-methylated organism, in contrast to most vertebrates and insects. Actually, *Drosophila* does methylate, but methylation is restricted to young embryos (Lyko 2001; Lyko et al. 2000), and mainly to cytosines associated with A and T nucleotides, which contrasts with the situation in the other organisms, in which methylation takes place on CpG dinucleotides. In addition, unlike other organisms, the methyltransferase responsible for methylation in *Drosophila* is Dnmt2, and the overall level of genome methylation is low. The impact of methylation in the *Drosophila* genome is still not well understood, but we can speculate that even if the levels of methylation are low, they could be sufficient to be involved in the regulation of TEs. Other insects such as the haplo/diploid honey bees (*Apis mellifera*) encompass a complete CpG methylation program (similar to humans) and might be a useful system to study epigenetic control of TEs in the future (Schaefer and Lyko 2007; Kucharski et al. 2008).

Global analysis of genome methylation rates are necessary to understand the real evolutionary impact of DNA methylation. High-performance capillary electrophoresis (HPCE) or liquid chromatography (LC) are common techniques to determine these global rates (Berdasco et al. 2009). Sequence information can be obtained by bisulfite techniques, which can be associated with methylation-sensitive restriction enzymes (Suzuki and Bird 2008, for a review). We can expect that in a near future methylomes from different species, but also from different individuals, will be available.

3.2.2
Histone Modifications

In the cell nucleus, DNA is associated with proteins (such as histones) to form the chromatin. Chemical modifications of the N-terminal as well as C-terminal histone tails will make the DNA more or less accessible to the transcription machinery. Thus, the post-translation state of the histones will regulate both gene activity (Buratowski and Moazed 2005) and TE activity (Hall et al. 2002; Tran et al. 2005). Histone modifications are considered to be reversible, sensitive cellular tools of gene regulation, able to control genes during development, turning them ON or OFF. The histone pattern of modifications thus represents the expression of genes and TEs in a genome (Gendrel et al. 2002; Goldberg et al. 2007; Martens et al. 2005), even if in *Drosophila*, the relationship with TE expression has still not been clearly demonstrated. If a TE is inserted in an open conformation of chromatin, invasion of the genome by this TE is possible. On the contrary, if the TE is inserted in closed chromatin, the transposition will be difficult and genome invasion less likely to occur.

3.2.3
RNA Interference (RNAi)

The existence of antisense RNAs produced by TEs have previously been reported and proposed to play a role in the control of their activity (for instance Lankenau et al. 1994 for the *micropia* LTR retrotransposon; Simmons et al. 1996 for the P transposon). Variability among strains in the efficiency of such a regulation was already proposed, underlying the need for a populational approach of these issues (Simmons et al. 1996).

More recently, small antisense RNA molecules have been demonstrated to play multiple roles in the regulation of gene expression (Fire et al. 1998). These include (1) the degradation of mRNA by small interfering RNA (siRNA), which leads to post-transcriptional repression of the genes (Post-Transcriptional Gene Silencing: PTGS), (2) repression of the translation of mRNA by micro RNA (miRNA) and (3) the repression of the transcription (Transcriptional Gene Silencing: TGS) (Buchon and Vaury 2006). This is a complex field of research, and various different pathways have been identified that lead to RNAi (Matranga and Zamore 2007). In *Drosophila*, a new type of siRNAs has just been described: "repeated-associated small interfering RNAs" or rasiRNAs. These entities are homologous to the antisense RNAs of some TE families (Aravin et al. 2007; Kavi et al. 2005; Pélisson et al. 2007; Vagin et al. 2004, 2006; Vaughn et al. 2007). This pathway involves Piwi-like proteins, members of the Argonaute family, not only in the germline but also in somatic cells (Kawamura et al. 2008).

The discovery of these rasiRNAs shed some significant light on the already identified control loci *flamenco* or COM, but whose precise molecular role was so far not understood. The heterochromatic regulatory locus COM, involved in the control of the *ZAM* and *Idefix* LTR retrotransposons, has indeed been demonstrated to direct TE somatic silencing through Piwi-dependent and -independent pathways (Desset et al. 2008). COM is located in the same region as the *flamenco* locus, which controls the *gypsy* LTR retrotransposon, also via the production of rasiRNAs (Pélisson et al. 2007). Brennecke et al. (2007) recently demonstrated that *flamenco* was a piRNA cluster, from which TE homologous piRNAs were produced. More recently, the P-M system and the I-R system have been shown to be entirely linked to the RNAi regulation (Brennecke et al. 2008).

In this regard, attention is paid to an intriguing electron-dense structure surrounding the nucleus of germline cells in *Drosophila*, called "nuage" (French word for "cloud"). Its exact function is still unknown, but it is now known that the molecular actors of the production of rasiRNAs localize into the nuage, which is therefore assumed to be a crucial structure in the regulation of TE expression (Lim and Kai 2007; Pane et al. 2007; see also Lankenau, vol. 3, this series). The germline-specific nuage of *Drosophila* seems to be conserved in other organisms. For instance, it corresponds to the chromatoid bodies of mammals, which also appears to be involved in the germline epigenetic control of TEs (Soper et al. 2008).

RNAi is implicated in a huge number of phenomena, and this must be integrated in the analysis of natural populations.

3.2.4
Population Epigenomics

It is becoming obvious that some epigenetic modifications are heritable (Cavalli and Paro 1999; Goldberg et al. 2007), although the mechanism by which this transmission is achieved remains unknown (Cuzin et al. 2008). The main challenge for geneticists is to understand the impact of this epigenetic code. Do different populations have different epigenetic pathways? What are the consequences of epigenetics for evolution? How does the genome take the changes in epigenetic states into account? Can we attribute differences between populations to differences in their epigenetic status? And finally, could differences in TE activity (transposition) be based on differing epigenetic controls? Very few studies have been done to analyze the epigenetic variability of TE regulation in natural populations, but some pioneer work has been performed in plants. In *Arabidopsis*, Vaughn and colleagues have recently shown that the cytosine methylation rate on genes differed depending on the ecotype analyzed, whereas no corresponding difference was detected in TE methylation (Vaughn et al. 2007). More recently, Zhai et al. (2008) have shown that these two ecotypes have different levels of siRNA production, implying that the siRNA cluster may not be the same in different populations. Similar studies have not yet been carried out in other organisms, although the data obtained from comparisons of the epigenome of monozygotic twins have shown how important the epigenome may be in determining the expression of the phenotype (Fraga et al. 2005).

Drosophila represents an excellent model for the study of population epigenomic variability, because of its relatively low – compared to some mammals or plants – number of TEs, the large number of mutant strains, and the availability of natural populations. Global methods for the analysis of the chromatin structure (Chip seq), for the analysis of the methylome and for the sequencing of small RNAs are now popular. Comparison of populations living in different environments and having different TE features (Vieira et al. 1999) is now easier, and does not even imply the artificial construction of the experiments. Population analysis of the epigenomes may be one of the ways used in the future to understand how individuals adapt to new environments, and shed new light on the study of genotype X environment interactions. In particular, some works have demonstrated the heritability of epigenetic marks (Cavalli and Paro 1999; Richards 2008).

4
Conclusion

Whole genome sequencing and population analyses should be viewed as complementary methods, each focusing on a different aspect of TE dynamics (see Fig. 3). Whole genome sequence analysis can be used to compare different families of TEs within a given genome, while population sequence analysis compares different genomes with regard to a given family of TEs.

Evolution is (predominantly) a stochastic process. The scale of impact of stochasticity depends on the constraints applied to a given section of the genome: the greater the constraints, the narrower the range of possible observable outcomes of evolution. This is one of the reasons why the variability observed for the TE load of a genome is much higher than that observed for the pool of protein coding genes (because weaker constraints act on TEs). This means that population-based studies can take into account the variability of pathways authorized by this combination of stochasticity and weaker constraints, whereas a single sequenced genome focuses on only one of the multitude of possible states.

The final alliance between sequenced genomes and population studies might be provided by the new sequencing techniques, such as 454 (Margulies et al. 2005), which will provide rapid sequencing of full genomes from different populations, and even from different individuals of a given species. The 454 technology, which is already very frequently used for the study of "non-model" organisms, might at first not appear to be ideal for the study

Y_1, Y_2, Y_3:
studied variables,
such as amount of TE,
copy number,
methylation level,etc.

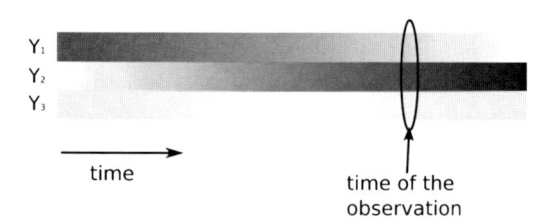

A. The sequenced genome

B. Populations

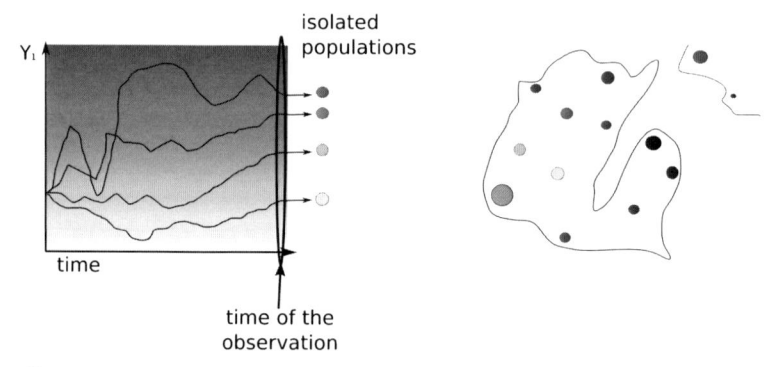

Fig. 3 Different, yet complementary, points of view provided by the study of the sequenced genome of a species (**A**), and several geographically distinct populations (**B**). The figure is drawn for any variable of study, called Y for convenience, such as for instance TE amount in the genome, copy number for a given TE family, level of DNA methylation in TE sequences, etc. Color variation from light to dark: variation of Y value from minimum to maximum. A, whole genome sequencing of an individual organism. The amount of data extracted is extensive since several Y variables (Y_1, Y_2, Y_3) can be studied at the same time. However, it provides only one value for each Y considered, without any information on their possible ranges of variation, contrary to populational studies. **B** *Left panel*: temporal evolution of the populations regarding to the value for Y_1. Several populations split from the original population, and start to evolve separately. With time, the values of Y_1 evolve differently in each population, so that at a given time of sampling, the populations are different regarding Y_1. **B** *Right panel*, spatial representation of populations and their corresponding values for Y_1. Isolated populations give rise to individual genomic samples. Each population displays a different value for Y_1

of repeated DNA, since the fragments sequenced are very short, but some studies have already been done and look promising (Macas et al. 2007).

Acknowledgements We are grateful to Christian Biémont for critical reading of the manuscript and to Monika Gosh for English correction. This work was supported by CNRS and Finovi.

References

Aminetzach YT, Macpherson JM, Petrov DA (2005) Pesticide resistance via transposition-mediated adaptive gene truncation in Drosophila. Science 309:764–767

Aravin AA, Hannon GJ, Brennecke J (2007) The Piwi-piRNA pathway provides an adaptive defense in the transposon arms race. Science 318:761–764

Bassett DE Jr, Basrai MA, Connelly C, Hyland KM, Kitagawa K, Mayer ML, Morrow DM, Page AM, Resto VA, Skibbens RV, Hieter P (1996) Exploiting the complete yeast genome sequence. Curr Opin Genet Dev 6:763–766

Begun DJ, Holloway AK, Stevens K, Hillier LW, Poh YP, Hahn MW, Nista PM, Jones CD et al (2007) Population genomics: whole-genome analysis of polymorphism and divergence in *Drosophila simulans*. PLoS Biol 5, e310

Bejerano G, Pheasant M, Makunin I, Stephen S, Kent WJ, Mattick JS, Haussler D (2004) Ultraconserved elements in the human genome. Science 304:1321–1325

Berdasco M, Fraga MF, Esteller M (2009) In: Tost J (ed) Quantification of global DNA methylation by capillary electrophoresis and mass spectrometry DNA methylation methods and protocols, 2nd edn. Springer Protocols, The Netherlands, pp 23–34

Biémont C, Monti-Dedieu L, Lemeunier F (2004) Detection of transposable elements in *Drosophila* salivary gland polytene chromosomes by in situ hybridization. Methods Mol Biol 260:21–28

Biémont C, Nardon C, Decelière G, Lepetit D, Loevenbruck C, Vieira C (2003) Worldwide distribution of transposable element copy number in natural populations of *Drosophila simulans*. Evolution 57:159–167

Biémont C, Tsitrone A, Vieira C, Hoogland C (1997) Transposable element distribution in *Drosophila*. Genetics 147:1997–1999

Biémont C (1994) Dynamic equilibrium between insertion and excision of P elements in highly inbred lines from an M' strain of *Drosophila melanogaster*. J Mol Evol 39:466–472

Biémont C, Lemeunier F, Garcia Guerreiro MP, Brookfield JF, Gautier C, Aulard S, Pasyukova EG (1994) Population dynamics of the copia, mdg1, mdg3, gypsy, and P transposable elements in natural populations of *Drosophila melanogaster*. Genet Res 63:197–212

Boissinot S, Furano AV (2005) The recent evolution of human L1 retrotransposons. Cytogenet Genome Res 110:402–406

Bowen NJ, McDonald JF (2001) *Drosophila* euchromatic LTR retrotransposons are much younger than the host species in which they reside. Genome Res 11:1527–1540

Brennecke J, Aravin AA, Stark A, Dus M, Kellis M, Sachidanandam R, Hannon GJ (2007) Discrete small RNA-generating loci as master regulators of transposon activity in *Drosophila*. Cell 128:1089–1103

Brennecke J, Malone CD, Aravin AA, Sachidanandam R, Stark A, Hannon GJ (2008) An epigenetic role for maternally inherited piRNAs in transposon silencing. Science 322:1387–1392

Bridges CB (1935) Salivary chromosome maps: with a key to the banding of the chromosomes of *Drosophila melanogaster*. J Hered 26:60–64

Buchon N, Vaury C (2006) RNAi: a defensive RNA-silencing against viruses and transposable elements. Heredity 96:195–202

Buratowski S, Moazed D (2005) Gene regulation: expression and silencing coupled. Nature 435:1174–1175

Cavalli G, Paro R (1999) Epigenetic inheritance of active chromatin after removal of the main transactivator. Science 286:955–958

Chan SW, Henderson IR, Jacobsen SE (2005) Gardening the genome: DNA methylation in Arabidopsis thaliana. Nat Rev Genet 6:590

Clark JB, Maddison WP, Kidwell MG (1994) Phylogenetic analysis supports horizontal transfer of P transposable elements. Mol Biol Evol 11:40–50

Conley AB, Miller WJ, Jordan IK (2008) Human cis natural antisense transcripts initiated by transposable elements. Trends Genet 24:53–56

Cordaux R, Hedges DJ, Herke SW, Batzer MA (2006a) Estimating the retrotransposition rate of human Alu elements. Gene 373:134–137

Cordaux R, Lee J, Dinoso L, Batzer MA (2006b) Recently integrated Alu retrotransposons are essentially neutral residents of the human genome. Gene 373:138–144

Costas J, Valade E, Naveira H (2001) Amplification and phylogenetic relationships of a subfamily of blood, a retrotransposable element of *Drosophila*. J Mol Evol 52:342–350

Crow JF, Simmons MJ (1983) The mutation load in *Drosophila*. In: Carson HL, Ashburner M, Thomson JN (eds) The genetics and biology of *Drosophila*. Academic Press, London, pp 1–35

Cuzin F, Grandjean V, Rassoulzadegan M (2008) Inherited variation at the epigenetic level: paramutation from the plant to the mouse. Curr Opin Genet Dev 18:193–196

Desset S, Buchon N, Meignin C, Coiffet M, Vaury C (2008) In *Drosophila melanogaster* the COM locus directs the somatic silencing of two retrotransposons through both Piwi-dependent and independent pathways. PLoS ONE 2:e1526

Drosophila 12 Genomes Consortium (2007) Evolution of genes and genomes on the *Drosophila* phylogeny. Nature 450:203–218

Eickbush TH, Furano AV (2002) Fruit flies and humans respond differently to retrotransposons. Curr Opin Genet Dev 12:669–674

Engels WR (1997) Invasions of P elements. Genetics 145:11–15

Engels WR, Preston SR (1980) Components of hybrid dysgenesis in a wild population of *Drosophila melanogaster*. Genetics 95:111–128

Fablet M, McDonald JF, Biémont C, Vieira C (2006) Ongoing loss of the tirant transposable element in natural populations of *Drosophila simulans*. Gene 375:54–62

Fablet M, Rebollo R, Biémont C, Vieira C (2007a) The evolution of retrotransposon regulatory regions and its consequences on the *Drosophila melanogaster* and *Homo sapiens* host genomes. Gene 390:84–91

Fablet M, Souames S, Biémont C, Vieira C (2007b) Evolutionary pathways of the tirant LTR retrotransposon in the *Drosophila melanogaster* subgroup of species. J Mol Evol 64:438–447

Fedoroff N, Botstein D (1992) The dynamic genomeBarbara McClintock's ideas in the century of genetics. Cold Spring Harbor Laboratory Press, Cold Spring Harbor, NY

Feschotte C, Pritham EJ (2007) DNA transposons and the evolution of eukaryotic genomes. Annu Rev Genet 41:331–368

Finnegan DJ (1989) Eukaryotic transposable elements and genome evolution. Trends Genet 5:103–107

Fire A, Xu S, Montgomery MK, Kostas SA, Driver SE, Mello CC (1998) Potent and specific genetic interference by double-stranded RNA in *Caenorhabditis elegans*. Nature 391:806–811

Fraga MF, Ballestar E, Paz MF, Ropero S, Setien F, Ballestar ML, Heine-Suner D, Cigudosa JC, Urioste M, Benitez J, Boix-Chornet M, Sanchez-Aguilera A, Ling C, Carlsson E,

Poulsen P, Vaag A, Stephan Z, Spector TD, Wu YZ, Plass C, Esteller M (2005) Epigenetic differences arise during the lifetime of monozygotic twins. Proc Natl Acad Sci USA 102:10604–10609

Gall JG, Pardue ML (1969) Formation and detection of RNA-DNA hybrid molecules in cytological preparations. Proc Natl Acad Sci USA 63:378–383

Gao X, Hou Y, Ebina H, Levin HL, Voytas DF (2008) Chromodomains direct integration of retrotransposons to heterochromatin. Genome Res 18:359–369

Gendrel AV, Lippman Z, Yordan C, Colot V, Martienssen RA (2002) Dependence of heterochromatic histone H3 methylation patterns on the *Arabidopsis* gene DDM1. Science 297:1871–1873

Goldberg AD, Allis CD, Bernstein E (2007) Epigenetics: a landscape takes shape. Cell 128:635–638

Goodstadt L, Ponting CP (2006) Phylogenetic reconstruction of orthology, paralogy, and conserved synteny for dog and human. PLoS Comput Biol 29:e133

Hall IM, Shankaranarayana GD, Noma K, Ayoub N, Cohen A, Grewal SI (2002) Establishment and maintenance of a heterochromatin domain. Science 297:2232-7

Hellman A, Chess A (2007) Gene body-specific methylation on the active X chromosome. Science 315:1141-3

Herrera RJ, Lowery RK, Alfonso A, McDonald JF, Luis JR (2006) Ancient retroviral insertions among human populations. J Hum Genet 51:353–362

Hiraizumi Y, Slatko B, Langley C, Nill A (1973) Recombination in *Drosophila melanogaster* male. Genetics 73:439–444

Holt RA, Subramanian GM, Halpern A, Sutton GG, Charlab R, Nusskern DR, Wincker P, Clark AG et al (2002) The genome sequence of the malaria mosquito *Anopheles gambiae*. Science 298:129–149

Honeybee Genome Sequencing Consortium (2006) Insights into social insects from the genome of the honeybee *Apis mellifera*. Nature 443:931–49

Hoogland C, Biémont C (1996) Chromosomal distribution of transposable elements in *Drosophila melanogaster*: test of the ectopic recombination model for maintenance of insertion site number. Genetics 144:197–204

Hoskins RA, Carlson JW, Kennedy C, Acevedo D, Evans-Holm M, Frise E, Wan KH, Park S, Mendez-Lago M, Rossi F, Villasante A, Dimitri P, Karpen GH, Celniker SE (2007) Sequence finishing and mapping of *Drosophila melanogaster* heterochromatin. Science 316:1625–1628

Human Genome Sequencing Consortium (2004) Finishing the euchromatic sequence of the human genome. Nature 431:931–945

Huijser P, Kirchhoff C, Lankenau DH, Hennig W (1988) Retrotransposon-like sequences are expressed in Y chromosomal lampbrush loops of *Drosophila hydei*. J Mol Biol 203:689–697

Jablonka E, Lamb MJ (2002) The changing concept of epigenetics. Ann NY Acad Sci 981:82–96

Jeltsch A, Nellen W, Lyko F (2006) Two substrates are better than one: dual specificities for Dnmt2 methyltransferases. Trends Biochem Sci 31:306–308

Jordan IK, McDonald JF (1999) Tempo and mode of Ty element evolution in *Saccharomyces cerevisiae*. Genetics 151:1341–1351

Jordan IK, McDonald JF (1998) Evidence for the role of recombination in the regulatory evolution of *Saccharomyces cerevisiae* Ty elements. J Mol Evol 47:14–20

Kalendar R, Tanskanen J, Immonen S, Nevo E, Schulman AH (2000) Genome evolution of wild barley (Hordeum spontaneum) by BARE-1 retrotransposon dynamics in response to sharp microclimatic divergence. Proc Natl Acad Sci USA 97:6603–6607

Kapitonov VV, Jurka J (2008) A universal classification of eukaryotic transposable elements implemented in Repbase. Nat Rev Genet 9:411–412; author reply 414

Kapitonov VV, Jurka J (2005) RAG1 core and V(D)J recombination signal sequences were derived from Transib transposons. PLoS Biol 3:e181

Kavi HH, Fernandez HR, Xie W, Birchler JA (2005) RNA silencing in *Drosophila*. FEBS Lett 579:5940–9

Kawamura Y, Saito K, Kin T, Ono Y, Asai K, Sunohara T, Okada TN, Siomi MC, Siomi H (2008) *Drosophila* endogenous small RNAs bind to Argonaute 2 in somatic cells. Nature 453:793–797

Kazazian HH Jr (1998) Mobile elements and disease. Curr Opin Genet Dev 8:343–350

Kidwell MG, Kidwell JF, Sved JA (1977) Hybrid dysgenesis in *Drosophila melanogaster*: a syndrome of aberrant traits, including mutation, sterility and male recombination. Genetics 86:813–833

Kidwell MG (1977) Reciprocal differences in female recombination associated with hybrid dysgenesis in *Drosophila melanogaster*. Genet Res 30:77–88

Kucharski R, Maleszka J, Foret S, Maleszka R (2008) Nutritional control of reproductive status in honeybees via DNA methylation. Science 319:1827–1830

Lander ES, Linton LM, Birren B, Nusbaum C, Zody MC, Baldwin J, Devon K, Dewar K et al. (2001) Initial sequencing and analysis of the human genome. Nature 409:860–921

Langley CH, Montgomery E, Hudson R, Kaplan N, Charlesworth B (1988) On the role of unequal exchange in the containment of transposable element copy number. Genet Res 52:223–235

Lankenau S, Corces VG, Lankenau DH (1994) The *Drosophila* micropia retrotransposon encodes a testis-specific antisense RNA complementary to reverse transcriptase. Mol Cell Biol 14:1764–1775

Lankenau DH, Huijser P, Jansen E, Miedema K, Hennig W (1990) DNA sequence comparison of micropia transposable elements from *Drosophila hydei* and *Drosophila melanogaster*. Chromosoma 99:111–117

Lankenau DH, Huijser P, Jansen E, Miedema K, Hennig W (1988) Micropia : a retrotransposon of *Drosophila* combining structural features of DNA viruses, retroviruses and non-viral transposable elements. J Mol Biol 204:233–246

Le Rouzic A, Decelière G (2005) Models of the population genetics of transposable elements. Genet Res 85:171–181

Lerat E, Capy P, Biémont C (2002a) The relative abundance of dinucleotides in transposable elements in five species. Mol Biol Evol 19:964–967

Lerat E, Capy P, Biémont C (2002b) Codon usage by transposable elements and their host genes in five species. J Mol Evol 54:625–637

Lerat E, Rizzon C, Biémont C (2003) Sequence divergence within transposable element families in the *Drosophila melanogaster* genome. Genome Res 13:1889–1896

Lim AK, Kai T (2007) Unique germ-line organelle, nuage, functions to repress selfish genetic elements in *Drosophila melanogaster*. Proc Natl Acad Sci USA 104:6714–6719

Lipatov M, Lenkov K, Petrov DA, Bergman CM (2005) Paucity of chimeric gene-transposable element transcripts in the *Drosophila melanogaster* genome. BMC Biol 3:24

Lyko F (2001) DNA methylation learns to fly. Trends Genet 17:169–72

Lyko F, Ramsahoye BH, Jaenisch R (2000) DNA methylation in *Drosophila melanogaster*. Nature 408:538–40

Macas J, Neumann P, Navratilova A (2007) Repetitive DNA in the pea (*Pisum sativum* L.) genome: comprehensive characterization using 454 sequencing and comparison to soybean and Medicago truncatula. BMC Genomics 8:427

Macpherson JM, González J, Witten DM, Davis JC, Rosenberg NA, Hirsh AE, Petrov DA (2008) Nonadaptive explanations for signatures of partial selective sweeps in *Drosophila*. Mol Biol Evol 25:1025–1042

Malik HS, Eickbush TH (2001) Phylogenetic analysis of ribonuclease H domains suggests a late, chimeric origin of LTR retrotransposable elements and retroviruses. Genome Res 11:1187–1197

Malik HS, Burke WD, Eickbush TH (1999) The age and evolution of non-LTR retrotransposable elements. Mol Biol Evol 16:793–805

Malik HS, Eickbush TH (1999) Modular evolution of the integrase domain in the Ty3/Gypsy class of LTR retrotransposons. J Virol 73:5186–5190

Margulies M, Egholm M, Altman WE, Attiya S, Bader JS, Bemben LA, Berka J, Braverman, et al. (2005) Genome sequencing in microfabricated high-density picolitre reactors. Nature 437:376–380

Marhold J, Rothe N, Pauli A, Mund C, Kuehle K, Brueckner B, Lyko F (2004) Conservation of DNA methylation in dipteran insects. Insect Mol Biol 13:117–123

Martens JH, O'Sullivan RJ, Braunschweig U, Opravil S, Radolf M, Steinlein P, Jenuwein T (2005) The profile of repeat-associated histone lysine methylation states in the mouse epigenome. EMBO J 24:800–812

Mathews LM, Chi SY, Greenberg N, Ovchinnikov I, Swergold GD (2003) Large differences between LINE-1 amplification rates in the human and chimpanzee lineages. Am J Hum Genet 72:739–748

Matranga C, Zamore PD (2007) Small silencing RNAs. Curr Biol 17:R789–93

Matyunina LV, Jordan IK, McDonald JF (1996) Naturally occurring variation in copia expression is due to both element (cis) and host (trans) regulatory variation. Proc Natl Acad Sci USA 93:7097–102

McClintock B (1984) The significance of responses of the genome to challenge. Science 226:792–801

McClintock B (1950) The origin and behavior of mutable loci in maize. Proc Natl Acad Sci USA 36:344–355

McCollum AM, Ganko EW, Barass PA, Rodriguez JM, McDonald JF (2002) Evidence for the adaptive significance of an LTR retrotransposon sequence in a *Drosophila* heterochromatic gene. BMC Evol Biol 2:5

McDonald JF, Matyunina LV, Wilson S, Jordan IK, Bowen NJ and Miller WJ (1997) LTR retrotransposons and the evolution of eukaryotic enhancers. Genetica 100:3–13

Mhanni AA, McGowan RA (2004) Global changes in genomic methylation levels during early development of the zebrafish embryo. Dev Genes Evol 214:412–417

Miller WJ, McDonald JF, Pinsker W (1997) Molecular domestication of mobile elements. Genetica 100:261–270

Morgan TH, Sturtevant AH, Muller HJ, Bridges CB (1915) The mechanisms of Mendelian heredity. H Holt and Co, New York

Mugnier N, Biémont C, Vieira C (2005) New regulatory regions of *Drosophila* 412 retrotransposable element generated by recombination. Mol Biol Evol 22:747–757

Mugnier N, Gueguen L, Vieira C, Biémont C (2008) The heterochromatic copies of the LTR retrotransposons as a record of the genomic events that have shaped the *Drosophila melanogaster* genome. Gene 411:87–93

Nene V, Wortman JR, Lawson D, Haas B, Kodira C, Tu ZJ, Loftus B, Xi Z et al (2007) Genome sequence of *Aedes aegypti*, a major arbovirus vector. Science 316:1718–1723

Nuzhdin SV, Mackay TFC (1995) The genomic rate of transposable elements movement in *Drosophila melanogaster*. Mol Biol Evol 12:180–181

Pane A, Wehr K, Schüpbach T (2007) *Zucchini* and *squash* encode two putative nucleases required for rasiRNA production in the *Drosophila* germline. Dev Cell 12:851–862

Pardue ML, Gall JG (1969) Molecular hybridization of radioactive DNA to the DNA of cytological preparations. Proc Natl Acad Sci USA 64:600–604

Peaston AE, Evsikov AV, Graber JH, de Vries WN, Holbrook AE, Solter D, Knowles BB (2004) Retrotransposons regulate host genes in mouse oocytes and preimplantation embryos. Dev Cell 7:597–606

Pélisson A, Sarot E, Payen-Groschene G, Bucheton A (2007) A novel repeat-associated small interfering RNA-mediated silencing pathway downregulates complementary sense gypsy transcripts in somatic cells of the *Drosophila* ovary. J Virol 81:1951–60

Picard G (1976) Non-mendelian female sterility in *Drosophila melanogaster*: hereditary transmission of the I factor. Genetics 1983:107–123

Ponger L, Li WH (2005) Evolutionary diversification of DNA methyltransferases in eukaryotic genomes. Mol Biol Evol 22:1119–1128

Pradhan S, Esteve PO (2003) Mammalian DNA (cytosine-5) methyltransferases and their expression. Clin Immunol 109:6–16

Rebollo R, Lerat E, Kleine Lopez L, Biémont C, Vieira C (2008) Losing *helena*: the extinction of a *Drosophila* LINE-like element. BMC Genomics 9:149

Reik W (2007) Stability and flexibility of epigenetic gene regulation in mammalian development. Nature 447:425–32

Richards EJ (2008) Population epigenetics. Curr Opin Genet Dev 18:221–226

Saedler H, Starlinger P (1992) Twenty-five years of transposable element research in Köln. In: Fedoroff N, Botstein D (eds) The dynamic genome. Cold Spring Harbor, Cold Spring Harbor, pp 243–263

SanMiguel P, Gaut BS, Tikhonov A, Nakajima Y, Bennetzen JL (1998) The paleontology of intergene retrotransposons of maize. Nat Genet 20:43–45

Schaefer M, Lyko F (2007) DNA methylation with a sting: an active DNA methylation system in the honeybee. Bioessays 29:208–211

Shapiro JA (1969) Mutations caused by the insertion of genetic material into the galactose operon of *Escherichia coli*. J Mol Biol 40:93–105

Silva JC, Kidwell MG (2004) Evolution of P elements in natural populations of *Drosophila willistoni* and *D. sturtevanti*. Genetics 168:1323–1335

Simmons MJ, Raymond JD, Grimes CD, Belinco C, Haake BC, Jordan M, Lund C, Ojala TA, Papermaster D (1996) Repression of hybrid dysgenesis in *Drosophila melanogaster* by heat-shock-inducible sense and antisense P-element constructs. Genetics 144:1529–1544

Slotkin RK, Martienssen R (2007) Transposable elements and the epigenetic regulation of the genome. Nat Rev Genet 8:272–285

Soper SFC, van der Heijden GW, Hardiman TC, Goodheart M, Martin SL, de Boer P, Bortvin A (2008) Mouse Maelstrom, a component of nuage, is essential for spermatogenesis and transposon repression in meiosis. Dev Cell 15:285–297

Suh DS, Choi EH, Yamazaki T, Harada K (1995) Studies on the transposition rates of mobile genetic elements in a natural population of *Drosophila melanogaster*. Mol Biol Evol 12:748–758

Suzuki MM, Bird A (2008) DNA methylation landscapes: provocative insights from epigenomics. Nat Rev Genetics 9:465–476

Taft JR, Mattick JS (2003) Increasing biological complexity is positively correlated with the relative genome-wide expansion of non-protein-coding sequences. Genome Biol 5:P1

Thompson JN Jr, Woodruff RC (1980) Increased mutation in crosses between geographically separated strains of *Drosophila melanogaster*. Proc Natl Acad Sci USA 77:1059–1062

Tran RK, Zilberman D, de Bustos C, Ditt RF, Henikoff JG, Lindroth AM, Delrow J, Boyle T, Kwong S, Bryson TD, Jacobsen SE, Henikoff S (2005) Chromatin and siRNA pathways cooperate to maintain DNA methylation of small transposable elements in Arabidopsis. Genome Biol 6:R90

Tribolium Genome Sequencing Consortium (2008) The genome of the model beetle and pest *Tribolium castaneum*. Nature 452:949–955

Vagin VV, Klenov MS, Kalmykova AI, Stolyarenko AD, Kotelnikov RN and Gvozdev VA (2004) The RNA interference proteins and vasa locus are involved in the silencing of retrotransposons in the female germline of *Drosophila melanogaster*. RNA Biol 1:54–8

Vagin VV, Sigova A, Li C, Seitz H, Gvozdev V, Zamore PD (2006) A distinct small RNA pathway silences selfish genetic elements in the germline. Science 313:320–4

Vaughn MW, Tanurdžić M, Lippman Z, Jiang H, Carrasquillo R, Rabinowicz PD, Dedhia N, McCombie WR, Agier N, Bulski A, Colot V, Doerge RW and Martienssen RA (2007) Epigenetic Natural Variation in *Arabidopsis thaliana*. PLoS Biol 5:e174

Vieira C, Biémont C (2004) Transposable element dynamics in two sibling species *Drosophila melanogaster* and *Drosophila simulans*. Genetica 120:115–123

Vieira C, Piganeau G, Biémont C (2000) High copy numbers of multiple transposable element families in an Australian population of *Drosophila simulans*. Genet Res 76:117–119

Vieira C, Lepetit D, Dumont S, Biémont C (1999) Wake up of transposable elements following *Drosophila simulans* worldwide colonization. Mol Biol Evol 16:1251–1255

Vieira C, Aubry P, Lepetit D, Biémont C (1998) A temperature cline in copy number for 412 but not roo/B104 retrotransposons in populations of *Drosophila simulans*. Proc Biol Sci 265:1161–1165

Vieira C, Biémont C (1997) Transposition rate of the 412 retrotransposable element is independent of copy number in natural populations of *Drosophila simulans*. Mol Biol Evol 14:185–188

Vieira C, Biémont C (1996) Geographical variation in insertion site number of retrotransposon 412 in *Drosophila simulans*. J Mol Evol 42:443–451

Volff JN (2006) Turning junk into gold: domestication of transposable elements and the creation of new genes in eukaryotes. Bioessays 28:913–922

Wang Y, Jorda M, Jones PL, Maleszka R, Ling X, Robertson HM, Mizzen CA, Peinado MA, Robinson GE (2006) Functional CpG methylation system in a social insect. Science 314:645–647

Weil C, Martienssen R (2008) Epigenetic interactions between transposons and genes: lessons from plants. Curr Opin Genet Dev 18:188–192

Wicker T, Sabot F, Hua-Van A, Bennetzen JL, Capy P, Chalhoub B, Flavell A, Leroy P, Morgante M, Panaud O, Paux E, SanMiguel P, Schulman AH (2007) A unified classification system for eukaryotic transposable elements. Nat Rev Genet 8:973–982

Wilson S, Matyunina LV, McDonald JF (1998) An enhancer region within the copia untranslated leader contains binding sites for *Drosophila* regulatory proteins. Gene 209:239–46

Xia Q, Zhou Z, Lu C, Cheng D, Dai F, Li B, Zhao P, Zha X, et al (2004) A draft sequence for the genome of the domesticated silkworm (*Bombyx mori*). Science 306:1937–1940

Zhai J, Liu J, Liu B, Li P, Meyers BC, Chen X, Cao X (2008) Small RNA-directed epigenetic natural variation in Arabidopsis thaliana. PLoS Genet 4:e1000056

Genome Dyn Stab (4)
D.-H. Lankenau, J.-N. Volff: Transposons and the Dynamic Genome
DOI 10.1007/7050_018/Published online: 11 August 2006
© Springer-Verlag Berlin Heidelberg 2006

Morphological Characters from the Genome: SINE Insertion Polymorphism and Phylogenies

Agnès Dettaï[1] (✉) · Jean-Nicolas Volff[2]

[1]Département Systématique et Evolution, Muséum National d'Histoire Naturelle, Paris, France
adettai@mnhn.fr

[2]Physiologische Chemie I, Biozentrum, University of Würzburg, Würzburg, Germany

Abstract For the last fifteen years, researchers have been using SINE (short interspersed elements; non-autonomous retroposons) insertion polymorphism as characters for phylogeny. Although the collection of these characters is much less straightforward and much more work intensive than for classical sequence data, they are subject to very little homoplasy, and therefore allow more reliable determination of the phylogeny of species. As reversions are very rare, and the ancestral state (absence of the insertion) is known, these characters are orientated a priori. They are also good markers for population genetics. Because of their almost complete lack of homoplasy, character conflict in these characters is a better indicator of incomplete lineage sorting and hybridization than other types of data, even for ancient divergences. Only a few examples of convergencies and reversions have been identified, and after looking through hundreds of characters; moreover, most instances of homoplasy are identifiable as such, so SINE insertion polymorphism can still be regarded as very high quality characters. Constant progress has been made through the years for the isolation of new SINEs as well as for the isolation of new insertion loci, both by bioinformatic methods and by benchwork. Numerous dedicated computer programs are available, and the newly sequenced complete genomes allow their full scale utilization. SINE insertion polymorphism data has proved its interest on complex phylogenetic problems where morphological and sequence data were not resolutive. The improvements in its portability encourage an enlargement of its application to new taxa, where it will provide novel and high quality phylogenetic information.

1
On the Importance of Getting the Phylogeny Right

Comparative genomics is coming of age. The recent availability of whole genome sequences allows more powerful detection of genes and better reconstruction of the evolution of their function. Yet the conclusions drawn from any set of comparative data, whether evolutionary or functional, are dependent on the hypothesis of relationships between the compared taxa. Identifying paralogous and orthologous genes (Fitch 1970) can be considerably simplified by having access to a reliable phylogeny of the species (Cotton 2005). Reliable phylogenies are also a necessity for other evolutionary studies, from biogeography to host–parasite co-evolution, to comparative ethology, and to developmental biology. While more and more phylogenetic relationships are known (Springer et al. 2004), many others remain uncertain, even between model organisms (see Chen et al. 2004 for an example). The number of extant organisms is a problem in itself (1.8 million species described to date; Tree of Life project http://tolweb.org/tree/), and many species have not yet been included in any phylogeny. While its results are central for the interpretation of most modern biology, phylogenetic inference is quite often a difficult task. The analysis of sequence data can be misleading because of the frequent presence of homoplasy: a similarity that is not inherited from a common ancestor, but obtained independently in two lineages, either by convergence or by reversion. Many publications have proposed corrections and models to compensate for differing rates of evolution, possibilities of back mutations and other processes leading to homoplasy. Contradiction can exist between morphological and molecular sequence data, but discrepancies among the trees inferred from different molecular datasets are also quite frequent.

This has led many researchers to turn towards what has been dubbed "rare genomic changes" (RGC, Rokas and Holland 2000; Springer et al. 2004). Obtaining data in this method is more work-intensive than for classical sequence data comparison (Shedlock et al. 2004), but there are reasons to believe that these characters (new spliceosomal introns, larger deletions in coding sequences, acquisition of new exons or whole genes, retroposon insertions, gene fusions etc.) are more trustworthy than the comparison of even a large number of point mutations. For RGC, in contrast to point mutations, the probability of obtaining the same character state twice independently (convergence), or to revert to a previous state (reversion), is low.

One very peculiar category of RGCs is the insertion of SINEs[1] (short interspersed elements). Although very recent data has challenged their previous image of "perfect characters", evidence suggests that these insertions are only very rarely subject to undetectable homoplasy (Shedlock et al. 2004; Salem et al. 2003a; Van de Lagemaat et al. 2005). In contrast to the other RGCs, SINE insertions are present in great numbers in many eukaryote genomes, providing corroboration to numerous nodes (points of separation of groups within a tree), whereas the other RGCs only support a node here or there. Moreover, these characters are the only ones among the RGC for which efficient recovery protocols have been available for more than ten years and which are improving steadily, even if they are still much heavier to implement than classical sequence data comparisons (Okada et al. 2004; Shedlock et al. 2004). This, and the elaborate exploration of character homology that is possible for each character (in contrast to sequence data), makes SINE insertions more similar to morphological characters than classical sequence characters. Yet the neutrality of SINE insertion characters leaves them generally independent from

---

[1] SINEs: short (80 to 400 base pairs), non-coding, non-automomous, class I mobile retroelements.

selective pressures and constraints, probably reducing considerably the risks of convergence.

While it is the polymorphism of single insertions that serves as character for the reconstruction of a phylogeny, the understanding of the nature of these elements and of the mechanisms of their retroposition are very important for determining their utility and the limits of their application.

2
SINEs

SINEs are non-autonomous class I mobile elements that retropose through a copy-and-paste mechanism involving the reverse transcription of an RNA intermediate (Weiner et al. 1986); their insertion in the genome generates flanking target site duplications of variable length (Fig. 1). When they are transcribed, it is by RNA polymerase III, and they do not encode any protein. For their retroposition, they rely on the endonuclease and reverse transcriptase encoded by specific autonomous retroelements, their partner LINEs (long interspersed elements; Esnault et al. 2000; Kajikawa and Okada 2002; Dewannieux et al. 2003). SINE/LINE pairs generally share a structural relationship under the form of a common $3'$ tail that is necessary for the recognition by the enzymes encoded by the LINE (Kajikawa and Okada 2002), although some SINEs like the primate *Alu* elements have a less specific, oligo(A) tail. All SINEs contain either a tRNA- (Sakamoto and Okada 1985; Endoh and Okada 1986; Okada 1991), a 5S rRNA- (Kapitonov and Jurka 2003a), or a 7SL-derived part (Weiner 1980; Ullu and Tschudi 1984), which in-

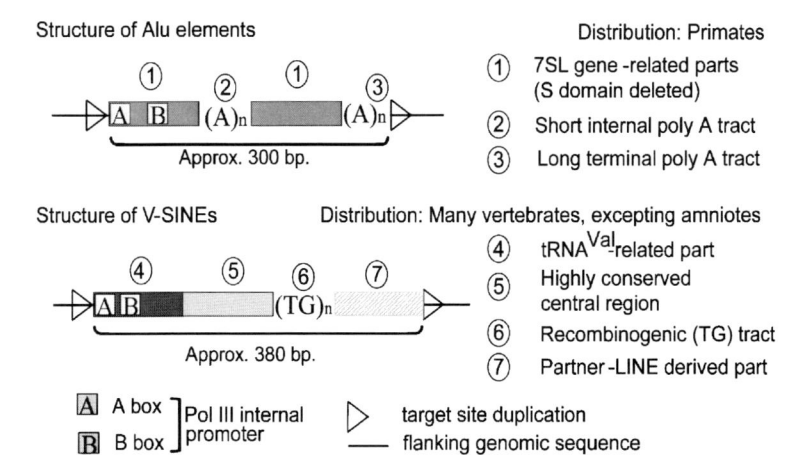

Fig. 1 Representation of two types of SINEs, the *Alu* family (Deininger et al. 1981; Quentin 1992; Dewannieux et al. 2003), and the V-SINEs superfamily (Ogiwara et al. 2000)

cludes a conserved polymerase III promoter, and often also a tRNA-unrelated region (Fig. 1). Most SINEs possess a tRNA-derived part, but currently only the zebrafish SINE3 elements are known to have a 5S-derived part (Kapitonov and Jurka 2003a). The presence of a 7SL-derived part is restricted to the rodent *B1* elements (Krayev et al. 1980), the primate *Alu* elements (Weiner 1980; Deininger et al. 1981), and the tree shrew type I and II elements (Nishihara et al. 2002). SINEs are widely distributed among Eukaryotes (Ohshima and Okada 2005), with a highly variable number of copies that can reach up to 10% of the whole genome (Lander et al. 2001; Batzer and Deininger 2002; Ogiwara et al. 2002). SINEs are completely absent from some species, for example *Drosophila* and the nematode *Caenorhabditis elegans* (Eickbush and Furano 2002). In addition to being present in vertebrates, they have been described in organisms as diverse as protozoans (like *Entamoeba histolytica,* van Dellen et al. 2002), cephalopod mollusks (Ohshima et al. 1993; Ohshima and Okada 1994), various insects (He et al. 1995; Feschotte et al. 2001; Tu 1999) and are widespread in plants (Schmidt 1999).

The activity of SINE families and subfamilies varies with time and lineages, like that of other mobile elements (Fig. 2 in Yang et al. 2004). They can show bursts of activity, and they can also go fully inactive and extinct, whether as a corollary of the extinction of their partner LINE like the MIR SINE/L2 LINE pair (Lander et al. 2001) or while their partner LINEs are still functional (Rinehart et al. 2004). Although they can be generally considered as neutral markers when integrated in non-functional regions of the genome, they can also have an impact on the genome by mediating deletions, duplications and translocations (Salem et al. 2003a, 2005) and cause disruptions by inserting into coding sequences (see Conley et al. 2005 for an example).

3
SINE Insertion Polymorphisms as Characters for Phylogeny

This method is based on the comparison of orthologous sites, some of which contain a SINE insertion, across individuals and species (Ryan and Dugaiczyk 1989; Batzer and Deininger 1991; Batzer et al. 1991; Okada 1991). Each insertion site is considered as a distinct character, with two possible character states: presence or absence of the SINE. Both states can be detected by a simple PCR with primers defined in the genomic sequences flanking the insertion, and confirmed by Southern blot hybridization and sequencing (Fig. 2). An important feature of the method is that it is possible to differentiate between absence of the insertion (shorter amplification product) and absence of the locus/technical problem (no amplification), thereby avoiding false negatives (Shedlock et al. 2000). SINE insertion polymorphisms (SINE IP) also have several other properties that distinguish them from other molecular characters.

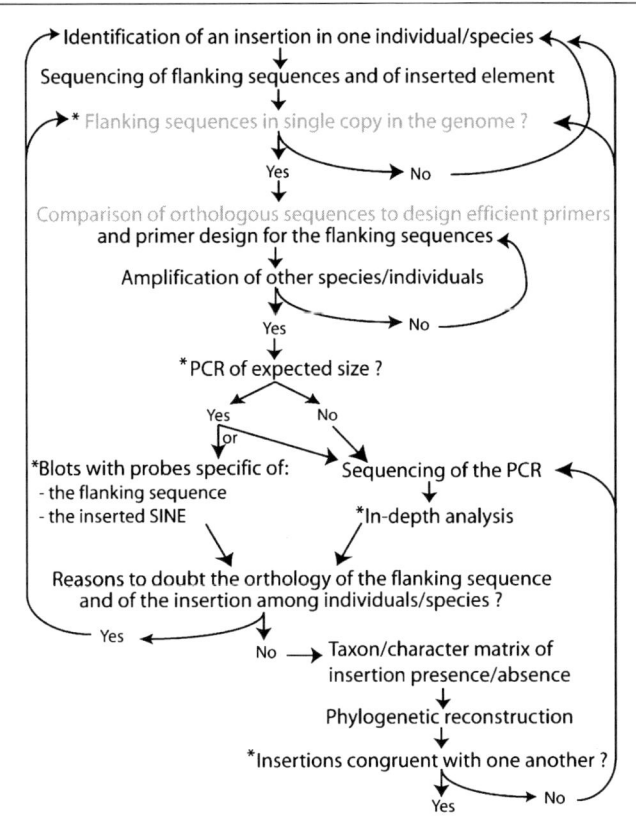

Fig. 2 Flowchart of a SINE insertion polymorphism study (Hillis 1999; Shedlock et al. 2000; Okada et al. 2004). * indicate orthology and homology tests of various stringencies; the steps common to all studies are in *black*, the steps depending on the amount of available data are in *grey*. See Okada et al. (2004) for a detailed, step-by-step description of the process

3.1
Character Quality vs. Character Quantity

As highlighted repeatedly in previous publications and reviews (Okada 1991; Cook and Tristem 1997; Hillis 1999; Miyamoto 1999; Shedlock and Okada 2000; Shedlock et al. 2000, 2004; Batzer and Deininger 2002, among others), SINE insertion polymorphism provides a source of high quality characters for phylogenetic and populational studies because of several unique properties. Each compared insertion locus is a single character and can support a single node, contrary to sequence data where a single sequence comparison provides many characters that support nodes at multiple levels in the tree. There is reason to believe that SINE insertion characters are largely not subject to convergence and reversion (back mutation to a state indistinguishable from the ancestral character state). A few cases of homoplasy have been described and will be discussed further in the paper. It follows from this that SINE insertion data will bring reliable information, which is especially valuable for groups where hypotheses of relationships already exist, but have been highly controversial (cichlids, Takahashi et al. 1998; mammals, Springer et al. 2004). Nonetheless, collecting SINE data is much more work-intensive than sequence data. Therefore, SINEs are less interesting when the aim of a study is to resolve from scratch the relationships of a group where nothing is known.

3.2
SINE Insertions are Apomorphies[2]

As established by Hennig (1950, 1965, 1966), only synapomorphies[3] permit the reconstruction of the phylogenetic relationships of taxa or individuals. But for most character types (sequence data, sequence insertions/deletions, and even most diagnostically valuable morphological data), apomorphies as a rule cannot be distinguished from plesiomorphies[4] a priori and researchers must resort to external information like an outgroup criterion to establish the polarity (orientation of the evolution) of the characters (Darlu and Tassy 1994; Kitching et al. 1998). The comparison of the character states present in the group of interest with the state observed within a related, but external taxon (or preferably, several) allows determination of which of the observed states is the ancestral one. When the closest outgroup is very distant, the evolutionary orientation is often problematic when using sequence and morphological data, as the characters are either too divergent to be meaningfully compared with the outgroup, or are too conserved to be informative in the studied group (i.e. symplesiomorphic[5]). SINE IP are exceptional in that regard, because the process is oriented: while new copies insert into the genome, excision is not a feature of these retroelements. Rare "excisions" (Van de Lagemaat et al. 2005) can probably be explained by endogenous double-strand break (DSB) repair mechanisms such as synthesis-dependent strand annealing (SDSA, Lankenau 1995; Lankenau and Gloor 1998). Therefore, insertions usually remain in place until they decay or are eliminated by larger deletions. As a consequence, the insertion-containing site is always an apomorphy, while the empty locus is the primitive ancestral state (Shedlock

---

[2] **Apomorphy:** a derived character state ("evolutionary novelty" for the group)

[3] **Synapomorphy:** an apomorphy shared by two or more groups, and that was already present in their last common ancestor. It can be used as evidence of the close relationship between the groups that present it.

[4] **Plesiomorphy:** an ancestral or primitive character state in the group of interest, but also present in some or all of the outgroups.

[5] **Symplesiomorphy:** an ancestral character state shared by at least two taxa. It does not give information on the relationships within the studied group, as the common ancestor and all the daughter lineages possess it.

et al. 2000). Unambiguous tree rooting is therefore possible in all cases, even when there are no closely related outgroups, or when it is not certain that the group of interest is monophyletic[6], i.e., comprises an ancestor and all its descendants (for instance Cheng et al. 2003).

3.3
Levels of Application

Depending on the period of activity of the SINE family that is studied, SINE IPs can provide informative characters for very recent events, making them useful for population genetics (see the review in Shedlock et al. 2004), as well as for much more ancient divergences. SINEs inserted in non-functional regions are generally considered as neutral markers and modeled accordingly (Shedlock et al. 2004). The spread of the allele containing the insertion to the whole diploid population, and the complete disappearance of the empty allele (fixation) under such conditions requires $4N$ generations ($2N$ for haploid populations), N being the population size (Hartl and Clark 1989). If speciation events from a same stem population (Fig. 3) are separated by a longer time, it is expected that all insertions that were present in a polymorphic state during the first speciation event will either have been fixed or have been lost (homozygous for the presence or absence of the insertion) by the time the next speciation event occurs. In such a case, all insertions should support the same groups without discordance (Fig. 3).

Fixed insertions are informative for phylogeny between species at various levels in the tree depending on the time at which the corresponding SINE family was active (Murata et al. 1993) (Fig. 4). Some examples of successful application of the SINE IP approach to phylogenetic inference are the intraordinal phylogeny of mammals (Shimamura et al. 1997; Nomura and Yasue 1999; Lum et al. 2000; Nikaido et al. 2003; Nishihara et al. 2005; Kriegs et al. 2006), the phylogeny within cetaceans (Nikaido et al. 2001, 2006), testudinoidean turtles (Sasaki et al. 2004), pecoran ruminants (Nijman et al. 2002), vespertilionid bats (Kawai et al. 2002), primates (Hamdi et al. 1999; Schmitz et al. 2001, 2004, 2005; Roos et al. 2004; Ray et al. 2005a; Xing et al. 2005), cichlids (Takahashi et al. 1998, 2001a,b; Sato et al. 2003; Terai et al. 2003, 2004), and crucifers (Tatout et al. 1999). While SINE IPs are informative for deeper phylogenies, the decay of the elements and the divergence of their flanking regions limit their application to events younger than 150–200 million years (Myrs), with a maximum efficiency being reached for 50 Myrs and less (Shedlock et al. 2004). As the similarity between the flanking sequences diminishes, the common origin of the insertion loci (orthology) is more difficult to ascertain, and more data will be missing because the amplification of the loci will

---

[6] **Monophyletic:** group encompassing all descendants of a stem group/species. The members of a monophyletic group are characterized by synapomorphic characters.

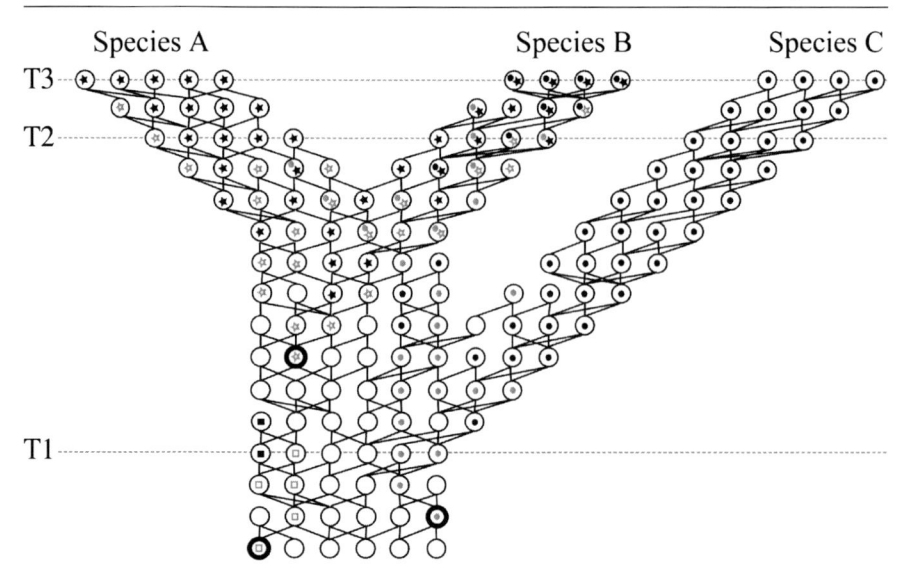

Species A   Species B   Species C

Fig. 3 Example of the history of three SINE insertions (☆, • and □) in a population before two close speciation events (modified from Hennig 1966). *Bold circles* indicate the
first heterozygous individual. *Grey symbols* indicate individuals heterozygous for an insertion; the same symbols in *black* indicate homozygous individuals for that locus. The
three *dashed lines* indicate three examples of SINE IP studies. *T1*: population genetics:
the informative characters are the polymorphic, not fixed alleles. *T2*: population genet-
ics/phylogenetics: the informative characters are the polymorphic, fixed and not fixed
alleles. *T3*: phylogenetics: the informative characters are the polymorphic, fixed alleles.
While the pattern of presence of the insertion ☆ reflects the true branching of species,
the distribution of insertion • is due to sorting of ancestral polymorphism. Insertion •
was not fixed when A/B diverged from C, and the locus was still polymorphic when A diverged from B. This insertion subsequently was lost by random genetic drift in species A,
while it was fixed in B and C. The incongruence between the groups supported by • (BC)
and by ☆ (AB) is indicative of the short time between the separation of species A/B from
species C and the separation of species A from B

be less successful (Shimamura et al. 1997; Shedlock et al. 2000). Problems can
start after a distance of a few million years, but defining primers within coding sequences that evolve more slowly allows efficient amplification for older
insertions (60 Myrs, Schmitz et al. 2001; 80 Myrs, Nishihara et al. 2005).

Non-fixed insertions[7] are useful markers for population genetics (Batzer
et al. 1991; Perna et al. 1992; T1, Fig. 3). They have been widely used in
humans (Batzer et al. 1994, 1995; Nei and Takezaki 1996; Stoneking et al.
1997; Sherry et al. 1997; Harpending et al. 1998; Hamdi et al. 1999; Jorde
et al. 2000; Comas et al. 2000, 2004; Watkins et al. 2001, 2003; Bamshad et al.

[7] **Non-fixed insertions**: insertions that are only present in some of the individuals of a species, and
that might either be fixed or eliminated by genetic drift.

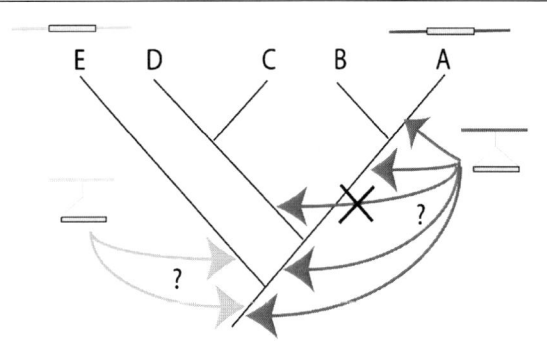

Fig. 4 Possible positions of the insertion event for a single locus in the tree, depending on the taxon where it is identified. Depending on the state of the locus for species B, C, D and E, a SINE insertion present in taxon A can support three different nodes, or is specific of A. SINE insertions present in taxon E can only support the monophyly of clade ABCDE, or be specific of E. Neither an insertion present in A, nor an insertion present in E can support clade CD, whatever the states in the other taxa are

2001, 2003; Carroll et al. 2001; Romualdi et al. 2002; Salem et al. 2003b; Vishwanathan et al. 2003; Maca-Meyer et al. 2004) but also in other species like *Aedes* mosquitoes (Barnes et al. 2005), baboons (Szmulewicz et al. 1999), and rice (Cheng et al. 2002, 2003). This type of data can also be used to infer population size (Stoneking et al. 1997; Sherry et al. 1997).

Non-fixed insertions can also provide information for recently diverged species (T2, Fig. 3), for instance for the elucidation of the relationships within salmonids (Murata et al. 1993, 1996, 1998; Hamada et al. 1998) and cichlids (Terai et al. 2004).

In many of these cases, the SINE IP method has successfully resolved some nodes where the classical morphological and molecular methods had failed to provide conclusive results. Moreover, SINE IPs can be considered as characters that are identical by descent only, because of the low probability of convergent insertions (see next section). This is especially important in cases where an ancestral polymorphism subsists over a long time and several speciation events, and different alleles are fixed in the daughter species independently of their order of separation (incomplete lineage sorting), or when interspecific hybridization occurs. In these cases, discrepancies between markers are the sign that the history of the species and the history of the molecular markers are not exactly the same. Sequence and microsatellite data can have a significant rate of homoplasy (Estoup and Cornuet 1999; Van Oppen et al. 2000), so disagreement among markers is not necessarily indicative of a complex history. On the contrary, SINE IPs are good indicators of shared ancestry and not largely subject to homoplasy (Shedlock et al. 2004), so disagreement among loci reliably indicates incomplete lineage sorting and interspecific hybridization (Takahashi et al. 2001, Fig. 3, T2, T3). Nonetheless, SINE characters can be homoplastic, although that is very rare. Incongruence due to a complex history can be distinguished from incongruence due to ho-moplasy by several characteristics. In the case of complex histories: first, the disagreement will be confined to particular nodes of the tree, those where the incomplete lineage sorting or hybridization events took place; and second, several loci will support each of a limited number of different alternatives (Shedlock et al. 2004; Nikaido et al. 2006). Homoplasy, on the other hand, is expected to be distributed more evenly on the tree, and among the taxa, therefore not presenting the restricted pattern that is observed.

3.3.1
Detecting Ancient Radiations

When several speciation events occur simultaneously or are only separated by a short time, no subgroup will be supported by character states present only in the group to the exclusion of all others – in other words, no character state will be perfectly and exclusively characteristic of a subgroup. The lack of sup-porting characters will yield an irresolution (polytomy) at the corresponding node in the tree. A multifurcation[8] in a tree can therefore indicate a simul-taneous speciation ("hard" polytomy). But while irresolution is frequent in phylogenetic trees, not all unresolved nodes indicate the existence of an ex-plosive radiation. An insufficient number of characters, multiple substitutions (Jackman et al. 1999), or contradictory gene trees due to various reconstruc-tion artifacts can also all lead to polytomies (Maddison 1989). This second type, dubbed "soft" polytomy, is due to our inability to uncover shared char-acters with the data used, and not to hard polytomies, the lack of shared characters due to simultaneous speciation events.

Demonstrating the existence of a hard polytomy is difficult. For recent events, it can be done by searching for traces of interspecific hybridization and incomplete lineage sorting (for example multiple sequence markers and microsatellite loci for a high number of individuals, Maddison and Knowles 2006). For more ancient events, subsequent mutations have obscured or erased altogether this signal in sequence data and microsatellite loci (Estoup and Cornuet 1999; Maddison and Knowles 2006). Analytical procedures and tests have been proposed to differentiate hard from soft polytomies on se-quence data (see for instance Jackman et al. 1999), but they rely on indirect evidence such as saturation levels, evaluation of structure etc. SINEs insertion data has two major advantages in this case. As explained previously, incon-gruent SINE insertion loci are a good indicator of incomplete lineage sorting and hybridization for recent events (Shedlock et al. 2004). The extreme rarity of reversions for SINE characters allows the conservation of this signal over time, and allows detection of more ancient incongruencies and, therefore,

---

[8] **Multifurcation:** also called polytomy or irresolution, part of a tree in which evolutionary rela-tionships cannot be fully resolved into dichotomies (more than two branches are diverging from a node).

of ancient simultaneous or neighboring speciations (Takahashi et al. 2001; Nikaido et al. 2006).

3.3.2
SINEs as Markers for the History of Genes

SINEs, especially older elements, are often recovered in gene-rich, GC-rich regions in the human genome (Lander et al. 2001), perhaps because they are preferentially retained in these regions (Li and Schmid 2001; Yang et al. 2004). Numerous examples of insertions within gene introns and exons have been described (Deininger and Batzer 1999). SINE insertion polymorphisms within related genes have been used with success to infer the phylogeny of gene families in primates. For example, shared insertions at paralogous loci have been studied (starting with Raisonnier 1991) and used with success to reconstruct the history of the major histocompatibility complex (MHC) DRB family of genes (Kriener et al. 2000), a multicopy subtelomeric region (Mefford et al. 2001), MHC class I duplicons (Kulski et al. 2000; Kulski and Dunn 2005), and S100A7 (psoriasin) gene duplications (Kulski et al. 2003).

3.4
Assessing Homology and Recognizing Homoplasy

The use of SINEs insertions as markers for phylogeny necessitates the identification of orthologous (Fitch 1970) insertions, i.e., insertions stemming from a single insertion event that occurred in a common ancestor.

The first requirement for the identification of orthologous insertions is the identification of the locus orthologous to the genomic locus carrying the insertion in different individuals (Fig. 2) and its selective amplification by PCR. This is done in the same way as for classical sequence markers, by sequence comparison between several species when available, and primer design to amplify selectively the sequences flanking the insertion. The pitfalls are the same as for classical sequence markers: the amplification of something other than the orthologous locus, either by lack of specificity of the primers, or because there are several similar loci in the genome. This last problem might be due to the insertion of the SINE in another mobile element (Batzer et al. 1991) or in a region that is duplicated in the genome (Luis et al. 2003). It can fully preclude the specific amplification of orthologous loci, so such insertion loci must not be used as characters for the phylogeny of species. The identification of multicopy loci should preferentially be performed before primer design (BLAST searches on sequence databanks to identify the flanking sequences for example as a transposable element). In cases where not enough sequences are available, one must rely on the detection of unexpected PCR patterns (Batzer et al. 2001) and sequence analysis for all the PCR products. Even when no homology problem is expected, the amplification of the correct locus is always controlled either by sequencing or by Southern blot hybridization of the amplified sequences with a labeled flanking sequence probe specific of the locus (Murata et al. 1993; Okada et al. 2004). If, indeed, the correct locus has been amplified, the flanking sequence probe will hybridize to the PCR bands, whether containing the insertion or not. It should be noted that paralogous loci with reasonable sequence similarity might also be positive using this method.

The second requirement for the identification of orthologous insertions, once the orthologous loci are identified, is the orthology of the insertion itself: that the SINEs observed at orthologous loci in different individuals/species stem from a single insertion event in a common ancestor. A Southern blot hybridization using the inserted SINE as a probe is the minimal control for this (Murata et al. 1993; Okada et al. 2004; Fig. 2). Sequencing of the PCR product and careful analysis of the sequences will make sure that the PCR size polymorphism is not due to the insertion of a similar element a few bases away (orthology of locus but not of insertion).

In both cases, sequencing and comparison of all the PCR is a more powerful control, allowing a more precise analysis of the flanking sequences, the inserted element, and the precise insertion site. Additionally, the obtained flanking sequences can be used as classical sequence markers to infer the phylogeny and divergence times of the species.

3.4.1
Convergence

Convergences are independently obtained similar derived states in different groups.

The study of SINE integration mechanisms has been of great importance in determining the probability of independent insertions at a same site. While such studies did not reveal strictly conserved nucleotide sequences at the integration sites except for an AATTTT preference for some SINEs (Jurka 1997), it has also been shown that integration is not fully random, with preferential integration in regions of particular composition (A + T rich regions: Jurka 1997; Tatout et al. 1998) or structure (Jurka et al. 1998; Tatout et al. 1998).

Several studies have uncovered cases of different retroposons inserted exactly at the same location in a gene. Two independent insertions of *Alu* elements were found between the same nucleotides in the factor IX gene (Wulff et al. 2000) and a germline *Alu* insertion exactly at the same site as a somatic LINE1 in the APC tumor suppressor gene (Halling et al. 1999). The BC200 RNA gene locus underwent two independent insertions, by a dimeric *Alu* element in Strepsirrhines primates and an *Alu* monomer-like element in anthropoids (Ludwig et al. 2005). Recently, a convergent disruptive insertion of two distinct elements (*Alu*Y and SVA) was reported in BTK, the gene responsible for X-linked agammaglobulinemia in two unrelated patients

(Conley et al. 2005). In Mys-9 (LTR retrotransposons specific to sigmodontine rodents), two hotspots within two very short sequences (248 and 150 bp) are present where at least six SINEs (B1, B2, and ID elements) and other elements (L1) have accumulated independently (Cantrell et al. 2001), including two B1 elements inserted exactly at the same position. In both studies (Cantrell et al. 2001; Conley et al. 2005), the target site duplications of the independent insertions are slightly different in length (16 versus 18bp and 9 versus 15bp respectively), and they can therefore also be distinguished on this basis. However, the convergent insertions found in the SMCY (select mouse cDNA on Y) gene in the wildcat (*Felix sylvestris*) and in the bobcat *Lynx rufus* (Pecon-Slattery et al. 2004) have completely indistinguishable flanking sequences.

Large-scale comparative studies have highlighted further cases. Six parallel insertions offset by a few base pairs were reported (Roy-Engel et al. 2002; Salem et al. 2003a; Xing et al. 2003). More recently, four examples of independent insertions at the same site, or a few base pairs away, in rhesus macaque and human or chimpanzee, and two independent insertions separated by 4 bp in human and gibbon have been discovered (Van de Lagemaat et al. 2005).

But while demonstrating that elements can, very rarely, insert twice exactly or almost exactly at the same location, these examples also show that most of these independent insertions can be differentiated through sequence comparison, or even Southern blot hybridization with a SINE-specific probe.

3.4.2
Reversion

Reversions are a return to a state indistinguishable from the primitive state. Until recently, no process that could excise a SINE precisely had been demonstrated, although in chimpanzees a single site identical to the empty (pre-insertion) site for an *Alu* insertion present in all other apes has been found (Hedges et al. 2004). But very recently, Van de Lagemaat and colleagues (2005) searched for all sites where a SINE could have been lost in the human or in the chimpanzee lineage, by looking for insertions present in rhesus macaque and one of the two apes, but absent in the other. Examination of deletions of *Alu* elements, of deletion internal to *Alus*, but also of genomic deletions of similar size (200–500 bp) showed an over-representation of deletions over the expected values when the sequences were flanked by identical repeats (including as short as 2 bp), leading them to conclude that the deletions might be partially due to a process of DSB repair with homologous pairing of the repeats. The DSB repair mechanism has been well defined (Lankenau and Gloor 1998). Such a process involving the target site duplications could lead to the exact excision of the element, and leave behind a perfect pre-insertion site. The proportion of such deletions over the total number of examined sites was small: 37 out of more than 7000 polymorphisms examined.

3.4.3
How much of a Problem is Homoplasy?

The identification of SINE deletions and independent insertions has required the analysis and comparison of hundreds or even thousands of insertion loci, which points to their rarity (Roy-Engel et al. 2002; Salem et al. 2003a; Van de Lagemaat et al. 2005). Furthermore, most of these cases can be spotted by the unexpected size of the amplification (because of a large deletion or the insertion of another type of sequence), no signal in the SINE-specific Southern blot hybridization (because of the insertion of a different mobile element), or at least by sequence differences revealing different (sub)families of elements or insertion positions. Without sequencing of the PCR product, an independent insertion of a similar element, even off-position by a few base pairs, could go undetected with only a PCR and two Southern blot hybridizations as controls.

But fortunately, even without access to all the sequences, an additional, even more powerful test is provided by the congruence among characters. If independent insertions at a same position are, as it seems, very rare events (Roy-Engel et al. 2002; Salem et al. 2005; Van de Lagemaat et al. 2005), the probability that several loci would be misleading in the same way is even lower. Until now, an almost perfect congruence has been observed among loci in the numerous phylogenetic studies using SINEs above the species level: more contradictions would be expected if a consequent number of observed insertions were in fact due to independent events. In the cases where contradictory data presented a significant fraction of the studied loci (Takahashi et al. 2001; Salem et al. 2003b; Terai et al. 2003, among others), incongruence was limited to one or a few nodes in the tree, and the lack of resolution or support by other types of data had already brought up the hypotheses of either interspecies hybridization or incomplete lineage sorting (Fig. 3). Nonetheless, this urges caution, especially in the cases where a single insertion supports a node, or where there is contradiction among loci at a node. A simple, first confirmatory step in these cases is to carefully compare the sequences, as most described independent insertions are still identifiable as such through this simple control.

3.4.4
Data Independence

When insertions are fixed within a species (and therefore are suitable for the study of divergences above species-level), any insertion event in the host genome can be considered to have inserted independently from any other (Shedlock et al. 2000), even when resulting from the same wave of retroposition (in opposition to Miyamoto 1999). The independence of insertion sites can be more of a problem when the insertions are not fixed (population studies). Loci physically close in the genome can be associated and fixed or lost together, and cannot be considered as independent support for a group. This is really a concern only for cases where the number of studied insertion sites is small and stochastic sampling problems can arise. The importance of measuring linkage disequilibrium of the loci (the degree of association existing between two variants of two loci) has long been recognized for microsatellite data, where it is routinely tested (Hartl and Clark 1989). These methods have been used for mixed SINE/microsatellite data (see for example Tishkoff et al. 2000).

4
Methods

4.1
Choice of Test Taxon

Whether the identification of the SINE and the single insertions will be performed *in silico* or by benchwork, the choice of the taxa on which this identification is performed is crucial. As only "filled" insertion sites can be identified, the "test" taxa must be part of the ingroup. If some phylogenetic information is available, choosing taxa that are not basal is better, as insertion polymorphism defined on a taxon that is sister-group to the rest cannot yield shared characters, or only shared characters supporting the whole group (Okada et al. 2004; Fig. 4).

4.2
Isolation of New SINEs

Many SINEs have already been described for various groups of organisms (Jurka et al. 2005; Ohshima and Okada 2005). However, if required, new elements can be identified through two possible approaches, the fast-developing computer-assisted identification, which strongly depends on the amount of sequence data available, and the classical benchwork one.

4.2.1
Computer-Assisted Identification

Previous to 1999, all *Drosophila* transposable elements reported had been discovered by benchwork (Kapitonov and Jurka 2003b). But, the recent abundance of sequence data allows for an efficient search for undescribed elements in the databanks for many organisms, and has been used with success in recent years (Churakov et al. 2005). Also, research on sequences specifically gathered for this purpose (Okada et al. 2004) have proved successful (Schmitz and Zischler 2003; Piskurek et al. 2003). Various tools have been developed, often with the primary objective of masking the repeated elements in order to ameliorate homology searches for coding sequences. The detection of SINEs is more complex than the search for elements containing coding genes because of their variable structure, and because of the sequence divergence between different copies of the same element.

The most widely used program for the identification of new elements is RECON (Bao and Eddy 2002), but numerous others have been proposed (RepeatFinder, Volfovsky et al. 2001; RepeatGluer, Pevzner et al. 2004). These methods start by making BLAST searches of the input sequence or genome versus itself, and then identify and classify the repetitive elements. Schematically, RECON uses the BLAST coordinates and the sharp changes in density of BLAST hits to a region to place boundaries between elements. Once related (alignable) elements are defined, it compares the aligned length to the total length of the elements and splits them into families when less than a certain proportion of the length can be aligned. Several other programs are available: a faster alternative to RECON (Price et al. 2005), and the implementation of a different method for repeat identification by focusing on the alignments that form a pattern characteristic of a given type of repeat (PILER, Edgar and Myers 2005). But these programs do not generate any repeat library, so additional analysis and grouping of the results are necessary.

An interesting new program, ReAS (recovery of ancestral sequences, Li et al. 2005), works directly on whole genome shotgun sequences and therefore avoids including artifacts that are due to possible misassemblies of the genome. Some problems must still be resolved manually, and this program is rather recommended as a starting point for the annotation of the complete mobile element content of a sequenced genome (Li et al. 2005).

SINEs can also be recovered by a genome-wide search for target site duplications separated by a given interval. For example, this approach allowed a new class of mammalian specific retropseudogenes to be uncovered (tailless SINEs, Schmitz et al. 2004).

In silico information about new elements, or about the description of a known element in a new species, requires experimental confirmation by Southern blot hybridization or PCR, as contamination problems and taxonomic misidentifications are commonplace in sequence databases (A. Dettaï et al., unpublished data).

4.2.2
Benchwork Identification

But, for many groups of organisms, little sequence data is available. In these cases, data must first be gathered by benchwork. Endoh and Okada (1986) proposed a method that is still widely used when no SINEs are known from the organisms of choice (Okada et al. 2004). It is based on in vitro transcription of genomic DNA in HeLa cell extracts. When one or several discrete bands are detected by electrophoresis, they are used as marked probe to screen genomic libraries of the studied species; the positive clones are subsequently sequenced. The complete protocol is described precisely (Okada et al. 2004), including the classification of SINEs in families and subfamilies, and the deduction of their tRNA-like structure. An alternative method to discover new SINEs was suggested in that paper and is the sequencing of at least 60 kb of sequence from the genome; the rationale is that SINEs are probably present in any larger stretch of sequence, as they often have a high copy number within a genome (this approach was used in Piskurek et al. 2003). Of course, this is dependent on the composition of the genome, some genomes being very SINE-poor (Eickbush and Furano 2002).

Borodulina and Kramerov (1999, 2005) suggested the use of short primers matching the two RNA polymerase III boxes common to all SINEs to isolate new elements by PCR.

Once new SINEs are isolated, the study of the level of sequence divergence between copies can serve as an indicator for the time-frame of activity of the family, and of the time window of species divergence where they might be informative (Okada et al. 2004).

The number of copies of the element and its distribution among taxa are also important factors to be evaluated, as the genomic behavior of a same SINE can be very variable in different species (Eickbush and Furano 2002; Ogiwara et al. 2002). This can be achieved by Southern blot or dot blot hybridization (Okada et al. 2004).

4.3
Isolation of New Insertion Loci

Once one or several elements are known for the group of organisms of interest, it is possible to proceed to the identification of insertion loci. Knowledge of the time-frame of activity of different families or subfamilies is a plus, as it allows selection for further studies of the elements that are the most likely to provide relevant information for the period of interest (Salem et al. 2003a; Xing et al. 2003).

4.3.1
In Silico

Once the existence of a SINE in a genome is detected, and its sequence is known, a search for new insertion loci can be started. When a large amount of sequence data is available for one species within the studied group, identifying new insertions is almost trivial, and can even be performed by simple BLAST searches using the sequence of the element as a query (Roy et al. 1999), although with a lower efficiency than with specialized programs. The most widely used program for repeat detection and annotation is Repeat-

Masker (A.F.A. Smit and P. Green unpublished), used with MaskerAid (Bedell et al. 2000). RepeatMasker screens sequences for low complexity DNA and interspersed repeats, using a precompiled library of repeats, RepBase Update (Jurka et al. 2005) or a user-defined custom library. The program outputs a version of the query sequence where all annotated repeats have been replaced by "Ns" (masked) as well as a detailed annotation of the repeats found in the submitted sequence. The flanking sequences of the masked elements can then be used to design primers. MaskerAid is a performance enhancement for RepeatMasker that can increase its speed more than 30 times (Bedell et al. 2000). This acceleration is especially meaningful since masking whole genomes has extensive time and computing power requirements. Repeat-Masker is available as an online version (http://www.repeatmasker.org/) that features premasked genomes, or can be downloaded for local use, with or without MaskerAid.

Another recent program that can be used is WindowMasker, a part of the NCBI C++ toolkit (Morgulis et al. 2006) that uses the input sequences directly and does not require a repeat library. Although it is described as much faster than RepeatMasker/MaskerAid, it has the great disadvantage of only masking the sequences without generating a list of the annotated repeats, so additional analysis is necessary to identify them and generate a repeat library. The authors have proposed two solutions to extract a library from the output: clustering the masked repeats according to their properties and reanalyzing them, or running RECON on well chosen subsets of masked sequence intervals (Morgulis et al. 2006).

These methods search for insertion sites, without any indication of whether they will be polymorphic, and therefore informative, at the studied level. Knowledge about the time-frame of activity of the various SINE families can help restricting the searches. For example, targeting the searches on young *Alu* elements from recently active subfamilies like Y (Batzer et al. 1996) has been used with success to detect polymorphic SINE insertions in humans (Roy-Engel et al. 2001; Xing et al. 2003; Salem et al. 2003a, among others). However, the polymorphism needs to be confirmed by flanking sequence PCR, and the number of informative sites is noticeably less than the number of tested sites (106 out of 475 in Carroll et al. 2001; 16 out of 250 in Terai et al. 2004).

More efficient approaches were recently proposed. They allow the *in silico* pre-selection of informative sites, but require the availability of a large amount of whole genome shotgun (Bennett et al. 2004) or of assembled sequences (Wang et al. 2006) for several individuals in human. Polymorphic insertion sites are detected by comparison of homologous sites in various individuals, and then can be further investigated in vitro. Large-scale application of this method with more than one million *Alu* insertions as a starting point yielded 800 polymorphic insertions, of which only one third had already been described, and was less biased towards young elements than previous studies (Wang et al. 2006). This type of approach could also be used for the identification of insertion polymorphisms for other SINEs, as well as for interspecific phylogenetic studies. Bashir and colleagues (2006) combined large amounts of sequence alignment for related species and RepeatMasker application to identify and analyze more than a 1000 insertion loci for mammalian phylogeny. Nevertheless the artifacts deriving from the method need to be manually sorted out.

4.3.2
In Vitro

In vitro identification of insertion loci has been widely used, and a precise description of the classical protocol is available (Okada et al. 2004). It relies on the identification of clones containing insertions from a genomic library using a labeled probe of a specific SINE in a Southern blot hybridization screening. The positive clones are then sequenced, PCR primers are designed for the sequences flanking the insertion, and other individuals/species are subsequently tested for the presence/absence of the insertion. But as only "filled" insertion sites can be detected by this method, the construction of a genomic bank as well as the clone screening must be repeated for several other species. Comparing the flanking sequences with sequence data from databanks can allow identification of multicopy flanking regions, help to design primers in more conserved areas, and lessen the risk of failed PCR (Fig. 2). Also, the method described by Borodulina and Kramerov (2005) can be used.

The recovery of flanking sequences can also be performed by methods like inverse PCR on genomic DNA digested by restriction enzymes not cutting in the SINE and subsequently circularized by ligation, with outward-facing primers specific of the element (Cheng et al. 2003; Ray et al. 2005b). Cloning and sequencing of the PCR products obtained follow this.

Several methods allow recovery by PCR of one side of the flanking sequences, with one oligonucleotide primer defined within the sequence of the SINE, and the other either a random 10 bp primer (described on L1-Ta LINEs, Sheen et al. 2000), an adapter-primer (linker PCR, as in Roy et al. 1999), or a suppression PCR[9] with first no second primer, and then a suppression adapter-specific primer (ATLAS method, also described on L1-Ta LINEs, Bagde et al. 2003). These methods allow specific selection for amplification of a subfamily that was active at the most interesting period for the projected study, as one of the primers is specific of the mobile element of interest. But, as only one flanking sequence is obtained by this method, it is only applicable in cases where a whole genome is available for closely related organisms, to

---

[9] **Suppression PCR:** based on the formation of a pan-like structure between the two suppression adapter sequences ligated to restricted genomic DNA.

determine the flanking sequence-specific primer for the other side. Polymorphism of the insertion is then determined by PCR with flanking primers, or by transposon display.

Recently, Mamedov and colleagues (2005) described a protocol that selects preferentially polymorphic sites with a high efficiency and sensitivity from a mix of human DNA from different individuals, and allows pinpointing SINE insertions present in even a single one of them in a heterozygous state. The approach includes a first step of whole genome selective amplification of *Alu* flanking sequences with *Alu*-specific primers and adapters performed on a mixture of genomes from several individuals. A second step selects the polymorphic sites by subtractive hybridization with genomic DNA from another individual. After cloning, one third of the clones were unique and contained insertions that proved polymorphic on the tested sampling. Numerous new polymorphisms were detected in this way, as many are absent from the few individuals represented in public genome databases (Mamedov et al. 2005; Wang et al. 2006). Such a method can be transposed to other species where little is available from the genome, as long as the sequence of the SINE is known.

4.4
Phylogenetic Reconstruction

Once insertion polymorphisms are established across the sampling, a taxon/character table can be prepared. Each insertion site is a single character, with two possible states: insertion present (longer PCR product) or insertion absent (shorter PCR product). When the whole locus was deleted, or the amplification/search for orthologous locus did not succeed, the character is coded as "?" for that species (Shedlock et al. 2000). In the absence of an applicable model of evolution for the insertions, an unweighted maximum parsimony search seems to be the best approach, as it can deal with missing data (Darlu and Tassy 1994; Kitching et al. 1998). As each insertion is a single two-state character and can only support a single node in the tree (Fig. 4), inclusion of as many insertion loci as possible is important. This also increases the probability of detecting incongruence among loci indicative of possible incomplete lineage sorting or hybridization.

The relatively small number of characters as well as their high quality makes the use of bootstrap questionable (Shedlock et al. 2004). The bootstrap method generates random datasets by sampling with replacement from the original sample, infers the tree(s) from each of these datasets, and calculates the proportion of the trees over all the generated datasets where a given group is present. Bootstrap values do not represent reliability of clades but their resistance to perturbations of the dataset (robustness), a very different notion (Darlu and Tassy 1994). Waddell and colleagues (2001) proposed a statistical test for clades specifically designed for SINE data in a likelihood framework that takes into account what is known about the evolution of the SINEs insertions.

4.5
Additional Information from Insertion Loci

As insertion loci are often sequenced to confirm their homology, the flanking sequence data that is generated can be successfully used to infer the phylogeny of the species (Lum et al. 2000; Nikaido et al. 2001), or to calculate lineage divergence time as with other sequence data (Nikaido et al. 2001).

4.6
Insertion Polymorphism of Other Mobile Elements for Phylogenetic Uses

Miniature inverted-repeat transposable elements (MITEs, non-autonomous DNA transposons) have also been proposed as indicators for phylogenetic reconstruction (Iszvàk et al. 1999). LINE elements have already been used in several studies (for example Vincent et al. 2003; Mateus Pereira et al. 2005; Kriegs et al. 2006), although their larger size makes them less practical for the amplification of the complete insertion using primers derived from the flanking regions. Bioinformatic approaches, on the other hand, do not suffer from this drawback, and have been used for other elements with as much success as for SINEs (Ho et al. 2005).

5
Conclusion

Whether it derives from morphology, sequence data comparison, or other molecular comparative methods like the SINE IP, each source of information has brought valuable insights and new hypotheses on the history of molecules, structures, and organisms. The use of SINE IP is radically different from classical sequence comparison because of the depth of analysis that is achievable in the evaluation of the homology of each single character. That explorability is much more akin to the scrutiny made possible by morphological characters. The availability of an a priori orientation of the evolution also makes SINE insertion characters special. This polarization offers a unique possibility for reliably confirming the position of a root, no matter how distant are the outgroups, thereby bypassing a problem common in phylogenetic inference. Sequence and morphological data can be problematic when the outgroup is distant, as the markers and characters are either too divergent to be meaningfully compared with the outgroup, or are to conserved to be informative in the studied group. SINEs insertion characters are not more sensitive than other types of characters to incomplete lineage sorting; on the contrary, their nature of "identical by descent"-only characters makes them a better tool to identify this type of event than microsatellites or sequence markers (Shedlock et al. 2004).

New kinds of characters for phylogenetic reconstruction go through two phases (Hillis 1999; Shedlock et al. 2000; Hagen 1999): one "honeymoon" phase, where they are touted as the perfect markers, and a critical phase following the realization that they also have drawbacks. For the moment, critics of the SINE IP method have been relatively mild (Luckett and Hong 1998) and partially based on misunderstandings (see Shedlock and Okada 2000; Shedlock et al. 2000, for a review). The accumulation of data on SINE evolution has brought only few hints of drawbacks, especially when compared with the limitations of sequence data discovered over the years. And yet, DNA sequences have remained an invaluable source of data for phylogenetic as well as for functional studies. Despite the fact that new data has added nuances to the affirmation that SINE insertions are completely homoplasy-free characters, wider scale studies have confirmed that homoplasy, whether convergence or reversion, is indeed very rare. SINE insertion polymorphism still appears to be a source of high quality phylogenetic information, even if additional studies are needed to determine the exact proportion of homoplasy in groups other than primates. Hence, analysis in new groups of organisms is worthwhile, especially for clades where other kinds of data fail to resolve important nodes. The full description of protocols is available, and numerous new and interesting methods have been published in recent years that propose innovative solutions for both benchwork and bioinformatic approaches. Although implementation is still heavier than for sequence data comparison, description of new SINEs for various taxa, high throughput sequencing, large amounts of available sequence data, and modern benchwork techniques make this approach applicable to a wide range of groups of organisms.

Acknowledgements Agnes Dettaï was supported by an Alexander von Humboldt postdoctoral Fellowship. The Volff group is funded by the Biofuture program of the German Bundesministerium für Bildung und Forschung (BMBF). We received helpful comments on this manuscript from Gael Lancelot, Matthieu Andro, and Julien Lorion.

References

Badge RM, Alisch RS, Moran JV (2003) ATLAS: A system to selectively identify human specific L1 insertions. Am J Human Genet 72:823–838

Bamshad M, Kivisild T, Watkins WS, Dixon ME, Ricker CE, Rao BB, Naidu JM, Prasad BVR, Reddy PG, Rasanayagam A, Papiha SS, Villems R, Carroll ML, Nguyen SV, Batzer MA, Jorde LB (2001) Genetic evidence on the origins of Indian caste populations. Genome Res 11:994–1004

Bamshad MJ, Wooding S, Watkins WS, Ostler C, Batzer MA, Jorde LB (2003) Human population genetic structure and inference of group membership. Am Human Genet 72:578–589

Barnes MJ, Lobo NF, Coulibaly MB, Sagnon NF, Costantini C, Besansky NJ (2005) SINE insertion polymorphism on the X chromosome differentiates Anopheles gambiae molecular forms. Insect Mol Biol 14(4):353–363

Bashir A, Ye C, Price AL, Bafna V (2006) Orthologous repeats and mammalian phylogenetic inference. Genome Res 15:998–1006

Batzer MA, Deininger PL (1991) A human-specific subfamily of Alu sequences. Genomics 9:481–487

Batzer MA, Deininger PL (2002) Alu repeats and human genomic diversity. Nature Reviews Genet 3:370–379

Batzer MA, Gudi VA, Mena C, Foltz DW, Herrera RJ, Deininger PL (1991) Amplification dynamics of Human-specific (HS) Alu family members. Nucleic Acids Res 19:3619–3623

Batzer MA, Stoneking M, Alegria-Hartman M, Bazan H, Kass DH, Shaikh TH, Novick GE, Ioannou PA, Scheer WD, Herrera RJ (1994) African origin of human-specific polymorphic Alu insertions. Proc Natl Acad Sci USA 91(25):12288–12292

Batzer MA, Rubin CM, Hellmann-Blumberg U, Alegria-Hartman M, Leeflang EP, Stern JD, Bazan HA, Shaikh TH, Deininger PL, Schmid CW (1995) Dispersion and insertion polymorphism in two small subfamilies of recently amplified human Alu repeats. J Mol Biol 247(3):418–427

Batzer MA, Deininger PL, Hellmann-Blumberg U, Jurka J, Labuda D, Rubin CM, Schmid CW, Zietkiewicz E, Zuckerkandl E (1996) Standardized nomenclature for Alu repeats. J Mol Evol 42:3–6

Bedell JA, Korf I, Gish W (2000) MaskerAid: a performance enhancement to RepeatMasker. Bioinformatics 16(11):1040–1041

Bennett EA, Coleman LE, Tsui C, Pittard WS, Devine SE (2004) Natural genetic variation caused by transposable elements in humans. Genetics 168(2):933–951

Borodulina OR, Kramerov DA (1999) Wide distribution of short interspersed elements among eukaryotic genomes. FEBS Lett 457(3):409–413

Borodulina OR, Kramerov DA (2005) PCR-based approach to SINE isolation: Simple and complex SINEs. Gene 349:197–205

Burton FH, Loeb DD, Edgell MH, Hutchison CA, III (1991) L1 gene conversion or samesite transposition. Mol Biol Evol 8:609–619

Cantrell MA, Filanoski BJ, Ingermann AR, Olsson K, DiLuglio N, Lister Z, Wichman HA (2001) An ancient retrovirus-like element contains hot spots for SINE insertion. Genetics 158(2):769–777

Carroll ML, Roy-Engel AM, Nguyen SV, Salem AH, Vogel E, Vincent B, Myers J, Ahmad Z, Nguyen L, Sammarco M, Watkins WS, Henke J, Makalowski W, Jorde LB, Deininger PL, Batzer MA (2001) Large-scale analysis of the Alu Ya5 and Yb8 subfamilies and their contribution to human genomic diversity. J Mol Biol 311(1):17–40

Chen WJ, Orti G, Meyer A (2004) Novel evolutionary relationship among four fish model systems. Trends Genet 20(9):424–431

Cheng C, Tsuchimoto S, Ohtsubo H, Ohtsubo E (2002) Evolutionary relationships among rice species with AA genome based on SINE insertion analysis. Genes Genet Syst 77(5):323–334

Cheng C, Motohashi R, Tsuchimoto S, Fukuta Y, Ohtsubo H, Ohtsubo E (2003) Polyphyletic origin of cultivated rice: based on the interspersion pattern of SINEs. Mol Biol Evol 20(1):67–75

Churakov G, Smit AF, Brosius J, Schmitz J (2005) A novel abundant family of retroposed

elements (DAS-SINEs) in the nine-banded armadillo (Dasypus novemcinctus). Mol
   Biol Evol 22(4):886–893

Comas D, Calafell F, Benchemsi N, Helal A, Lefranc G, Stoneking M, Batzer MA, Bertranpetit J, Sajantila A (2000) Alu insertion polymorphisms in NW Africa and the Iberian
   Peninsula: evidence for a strong genetic boundary through the Gibraltar Straits. Human Genet 107(4):312–319

Comas D, Schmid H, Braeuer S, Flaiz C, Busquets A, Calafell F, Bertranpetit J, Scheil HG,
   Huckenbeck W, Efremovska L, Schmidt H (2004) Alu insertion polymorphisms in the
   Balkans and the origins of the Aromuns. Annals Human Genet 68(2):120–127

Conley ME, Partain JD, Norland SM, Shurtleff SA, Kazazian HH Jr (2005) Two independent retrotransposon insertions at the same site within the coding region of BTK.
   Hum Mutat 25(3):324–325

Cook J, Tristem M (1997) 'SINEs of the times' – transposable elements as clade markers
   for their hosts. TREE 12(8):295–297

Cotton JA (2005) Analytical methods for detecting paralogy in molecular datasets.
   Methods Enzym 395:700–724

Darlu P, Tassy P (1994) La reconstruction phylogénétique. http://lis.snv.jussieu.fr/sfs/
   publications_sfs.shtml

Deininger PL, Jolly DJ, Rubin CM, Friedmann T, Schmid CW (1981) Base sequence study
   of 300 nucleotide renatured repeated human DNA clones. J Mol Biol 151:17–33

Deininger PL, Batzer AM (1999) Alu repeats and human disease. Mol Genet Metabol
   67:183–193

Van Dellen K, Field J, Wang Z, Loftus B, Samuelson J (2002) LINEs and SINE-like elements
   of the protist Entamoeba histolytica. Gene 297(1–2):229–239

Dewannieux M, Esnault C, Heidmann T (2003) LINE-mediated retrotransposition of
   marked Alu sequences. Nat Genet 35:41–48

Bao Z, Eddy SR (2002) Automated de novo identification of repeat sequence families in
   sequenced genomes. Genome Res 12(8):1269–1276

Edgar RC, Myers EW (2005) PILER: identification and classification of genomic repeats.
   Bioinformatics 21(S1):i152–i158

Eickbush TH, Furano AV (2002) Fruit flies and humans respond differently to retrotransposons. Curr Opin Genet Dev 12(6):669–674

Endoh H, Okada N (1986) Total DNA transcription in vitro: a procedure to detect highly
   repetitive and transcribable sequences with tRNA-like structures. Proc Natl Acad
   Sci USA 83(2):251–255

Esnault C, Maestre J, Heidmann T (200) Human LINE retrotransposons generate processed pseudogenes. Nat Genet 24(4):363–367

Estoup A, Cornuet JM (1999) Microsatellite evolution: inferences from population data.
   In: Goldstein DB, Schlotterer C (eds) Microsatellites: evolution and applications. Oxford University Press, Oxford

Feschotte C, Fourrier N, Desmons I, Mouches C (2001) Birth of a retroposon: the
   Twin SINE family from the vector mosquito Culex pipiens may have originated from
   a dimeric tRNA precursor. Mol Biol Evol 18(1):74–84

Fitch WM (1970) Distinguishing homologous from analogous proteins. Syst Zool
   19(2):99–113

Halling KC, Lazzaro CR, Honchel R, Bufill JA, Powell SM, Arndt CA, Lindor NM (1999)
   hereditary desmoid disease in a family with a germline Alu I repeat mutation of the
   APC gene. Hum Hered 49:97–102

Hamada M, Himberg M, Bodaly RA, Reist JD, Okada N (1998) Monophyletic origin of the genera Stenodus and Coregonus as inferred from an analysis of the insertion of SINEs (short interspersed repetitive elements). Adv Limnol 50:383–389

Hamdi H, Nishio H, Zielinski R, Dugaiczyk A (1999) Origin and phylogenetic distribution of Alu DNA repeats: irreversible events in the evolution of primates. J Mol Biol 289(4):861–871

Hartl DL, Clark AG (1989) Principles of population genetics. Sinauer, Sunderland Massachussets

Harpending HC, Batzer MA, Gurven M, Jorde LB, Rogers AR, Sherry ST (1998) Genetic traces of ancient demography. Proc Natl Acad Sci USA 95:1961 1967

He H, Rovira C, Recco-Pimentel S, Liao C, Edstrom JE (1995) Polymorphic SINEs in chironomids with DNA derived from the R2 insertion site. J Mol Biol 245(1):34–42

Hedges DJ, Callinan PA, Cordaux R, Xing J, Barnes E, Batzer MA (2004) Differential Alu mobilization and polymorphism among the human and chimpanzee lineages. Genome Res 14:1068–1075

Hennig W (1950) Grundzüge einer Theorie der phylogenetischen Systematik. Deutscher Zentralverlag, Berlin

Hennig W (1965) Phylogenetic systematics. Ann Rev Entomol 10:97–116

Hennig W (1966) Phylogenetic systematics. University of Illinois Press, Illinois

Hillis DM (1999) SINEs of the perfect character. Proc Natl Acad Sci USA 96(18):9979–9981

Ho HJ, Ray DA, Salem AH, Myers JS, Batzer MA (2005) Straightening out the LINEs: LINE-1 element orthologous loci. Genomics 85:201–207

Izsvàk Z, Ivics Z, Shimoda N, Mohn D, Okamoto H, Hackett PB (1999) Short inverted-repeat transposable elements in teleost fish and implications for a mechanism of their amplification. J Mol Evol 48(1):13–21

Jackman TR, Larson A, De Queiroz K, Losos JB (1999) Phylogenetic relationships and tempo of early diversification in Anolis lizards. Syst Biol 48(2):254–285

Jorde LB, Watkins WS, Bamshad MJ, Dixon ME, Ricker CE, Seielstad MT, Batzer MA (2000) The distribution of human genetic diversity: a comparison of mitochondrial, autosomal, and Y chromosome data. Am J Human Genet 66:979–988

Jurka J (1997) Sequence patterns indicate an enzymatic involvement in integration of mammalian retroposons. Proc Natl Acad Sci USA 94:1872–1877

Jurka J (1998) Repeats in genomic DNA: mining and meaning. Curr Opin Struct Biol 8:333–337

Jurka J, Kapitonov VV, Pavlicek A, Klonowski P, Kohany O, Walichiewicz J (2005) Repbase update, a database of eukaryotic repetitive elements. Cytogenet Genome Res 110:462–467

Kajikawa M, Okada N (2002) LINEs mobilize SINEs in the eel through a shared 3' sequence. Cell 111(3):433–444

Kapitonov VV, Jurka J (2003a) A novel class of SINE elements derived from 5S rRNA. Mol Biol Evol 20:694–702

Kapitonov VV, Jurka J (2003b) Molecular paleontology of transposable elements in the Drosophila melanogaster genome. Proc Natl Acad Sci USA 100(11):6569–6574

Kawai K, Nikaido M, Harada M, Matsumura S, Lin LK, Wu Y, Hasegawa M, Okada N (2002) Intra- and interfamily relationships of Vespertilionidae inferred by various molecular markers including SINE insertion data. J Mol Evol 55(3):284–301

Kitching IJ, Forey PL, Humphries JH, Williams DM (1998) Cladistics – the theory and practice of parsimony analysis. The Systematics Association Publication n°11, Oxford University Press, New York

Krayev AS, Kramerov DA, Skryabin KG, Ryskov AP, Bayev AA, Georgiev GP (1980) The nucleotide sequence of the ubiquitous repetitive DNA sequence B1 complementary to the most abundant class of mouse fold-back RNA. Nucleic Acids Res 8(6):1201–1215

Kriegs JO, Churakov G, Kiefmann M, Jordan U, Brosius J, Schmitz J (2006) Retroposed elements as archives for the evolutionary history of placental mammals. PLoS Biol 4(4):e91

Kriener K, O'h Uigin C, Klein J (2000) Alu elements support independent origin of prosimian, platyrrhine, and catarrhine Mhc-DRB genes. Genome Res 10(5):634–643

Kulski JK, Dunn DS (2005) Polymorphic Alu insertions within the major histocompatibility complex class I genomic region: a brief review. Cytogenet Genome Res 110(1–4):193–202

Kulski JK, Gaudieri S, Dawkins RL (2000) Using alu J elements as molecular clocks to trace the evolutionary relationships between duplicated HLA class I genomic segments. J Mol Evol 50(6):510–519

Kulski JK, Lim CP, Dunn DS, Bellgard M (2003) Genomic and phylogenetic analysis of the S100A7 (Psoriasin) gene duplications within the region of the S100 gene cluster on human chromosome 1q21. J Mol Evol 56(4):397–406

Lander ES, International Human Genome Sequencing Consortium et al. (2001) Initial sequencing and analysis of the human genome. Nature 409(6822):860–921

Lankenau D-H (1995) Genetics of genetics in Drosophila: P elements serving the study of homologous recombination, gene conversion and targeting. Chromosoma 103:659–668

Lankenau D-H, Gloor GB (1998) In vivo gap repair in Drosophila: a one-way street with many destinations. BioEssays 20:317–327

Li TH, Schmid CW (2001) Differential stress induction of individual Alu loci: implications for transcription and retrotransposition. Gene 276(1–2):135–141

Li R, Ye J, Li S, Wang J, Han Y, Ye C, Wang J, Yang H, Yu J, Wong GK, Wang J (2005) ReAS: Recovery of ancestral sequences for transposable elements from the unassembled reads of a whole genome shotgun. PLoS Comput Biol 1(4):e43

Luckett P, Hong N (1998) Phylogenetic relationships between the orders Artiodactyla and Cetacea: a combined assessment of morphological and molecular evidence. J Mammal Evol 5:127–182

Ludwig A, Rozhdestvensky TS, Kuryshev VY, Schmitz J, Brosius J (2005) An unusual primate locus that attracted two independent Alu insertions and facilitates their transcription. J Mol Biol 350(2):200–214

Luis JR, Terreros MC, Martinez L, Rojas D, Herrera RJ (2003) Two problematic human polymorphic Alu insertions. Electrophoresis 24(14):2290–2294

Lum JK, Nikaido M, Shimamura M, Shimodaira H, Shedlock AM, Okada N, Hasegawa M (2000) Consistency of SINE insertion topology and flanking sequence tree: Quantifying relationships among cetartiodactyls. Mol Biol Evol 17(10):1417–1424

Maca-Meyer N, Villar J, Perez-Mendez L, Cabrera de Leon A, Flores C (2004) A tale of aborigines, conquerors and slaves: Alu insertion polymorphisms and the peopling of Canary Islands. Annals Human Genet 68(6):600–605

Maddison WP (1989) Reconstructing character evoution on polytomous cladograms. Cladistics 5:365–377

Maddison WP, Knowles LL (2006) Inferring phylogeny despite incomplete lineage sorting. Syst Biol 55(1):21–30

Mamedov IZ, Arzumanyan ES, Amosova AL, Lebedev YB, Sverdlov ED (2005) Whole-genome experimental identification of insertion/deletion polymorphisms of interspersed repeats by a new general approach. Nucleic Acids Res 33(2):e16

Mateus Pereira LH, Socorro A, Fernandez I, Masleh M, Vidal D, Bianchi NO, Bonatto SL, Salzano FM, Herrera RJ (2005) Phylogenetic information in polymorphic L1 and Alu insertions from East Asians and Native American populations. Am J Phys Anthropology 128(1):171–184

Mefford HC, Linardopoulou E, Coil D, van den Engh G, Trask BJ (2001) Comparative sequencing of a multicopy subtelomeric region containing olfactory receptor genes reveals multiple interactions between non-homologous chromosomes. Hum Mol Genet 10(21):2363–2372

Miyamoto MM (1999) Molecular systematics: perfect SINEs of evolutionary history? Curr Biol 9(21):R816–R819

Morgulis A, Gertz EM, Schaffer AA, Agarwala R (2006) WindowMasker: window-based masker for sequenced genomes. Bioinformatics 22(2):134–141

Murata S, Takasaki N, Saitoh M, Okada N (1993) Determination of the phylogenetic relationships among Pacific salmonids by using short interspersed elements (SINEs) as temporal landmarks of evolution. Proc Natl Acad Sci USA 90(15):6995–6999

Murata S, Takasaki N, Saitoh M, Tachida H, Okada N (1996) Details of retropositional genome dynamics that provide a rationale for a generic division: the distinct branching of all the Pacific salmon and trout (Oncorhynchus) from the Atlantic salmon and trout (Salmo). Genetics 142(3):915–926

Murata S, Takasaki N, Okazaki T, Kobayashi T, Numachi K, Chang K-H, Okada N (1998) Molecular evidence from short interspersed elements (SINEs) that Oncorhynchus masou (cherry salmon) is monophyletic. Canadian J Fisheries Aqua Sci 55(8):1864–1870

Nei M, Takezaki N (1996) The root of the phylogenetic tree of human populations. Mol Biol Evol 13(1):170–177

Nijman IJ, van Tessel P, Lenstra JA (2002) SINE retrotransposition during the evolution of the pecoran ruminants. J Mol Evol 54(1):9–16

Nikaido M, Matsuno F, Hamilton H, Brownell RL, Cao Y, Ding W, Zuoyan Z, Shedlock AM, Fordyce RE, Hasegawa M, Okada N (2001) Retroposon analysis of major cetacean lineages: the monophyly of toothed whales and the paraphyly of river dolphins. Proc Natl Acad Sci USA 98(13):7384–7389

Nikaido M, Nishihara H, Hukumoto Y, Okada N (2003) Ancient SINEs from African endemic mammals. Mol Biol Evol 20(4):522–527

Nikaido M, Hamilton H, Makino H, Sasaki T, Takahashi K, Goto M, Kanda N, Pastene LA, Okada N (2006) Baleen whale phylogeny and a past extensive radiation event revealed by SINE insertion analysis. Mol Biol Evol 23(5):866–873

Nishihara H, Terai Y, Okada N (2002) Characterization of novel Alu- and tRNA-related SINEs from the tree shrew and evolutionary implications of their origins. Mol Biol Evol 19(11):1964–1972

Nishihara H, Satta Y, Nikaido M, Thewissen JG, Stanhope MJ, Okada N (2005) A retroposon analysis of Afrotherian phylogeny. Mol Biol Evol 22(9):1823–1833

Nomura O, Yasue H (1999) Genetic relationships among hippopotamus, whales, and bovine based on SINE insertion analysis. Mamm Genome 10(5):526–527

Ogiwara I, Miya M, Ohshima K, Okada N (2002) V-SINEs: a new superfamily of vertebrate SINEs that are widespread in vertebrate genomes and retain a strongly conserved segment within each repetitive unit. Genome Res 12(2):316–324

Ohshima K, Okada N (1994) Generality of the tRNA origin of short interspersed repetitive elements (SINEs). Characterization of three different tRNA-derived retroposons in the octopus. J Mol Biol 243(1):25–37

Ohshima K, Okada N (2005) SINEs and LINEs: symbionts of eukaryotic genomes with a common tail. Cytogenet Genome Res 110(1–4):475–490

Ohshima K, Koishi R, Matsuo M, Okada N (1993) Several short interspersed repetitive elements (SINEs) in distant species may have originated from a common ancestral retrovirus: characterization of a squid SINE and a possible mechanism for generation of tRNA-derived retroposons. Proc Natl Acad Sci USA 90(13):6260–6264

Okada N (1991) SINESs: short interspersed repeated elements of the eukaryotic genome. TREE 6(11):358–361

Okada N, Shedlock AM, Nikaido M (2004) Retroposon mapping in molecular systematics. Methods Mol Biol 260:189–226

Perna NT, Batzer MA, Deininger PL, Stoneking M (1992) Alu insertion polymorphism: A new type of marker for human population studies. Human Biol 64:641–648

Pevzner PA, Tang H, Tesler G (2003) De novo repeat classification and fragment assembly. Genome Res 14(9):1786–1796

Piskurek O, Nikaido M, Boeadi, Baba M, Okada N (2003) Unique mammalian tRNA-derived repetitive elements in dermopterans: the t-SINE family and its retrotransposition through multiple sources. Mol Biol Evol 20(10):1659–1668

Price AL, Jones NC, Pevzner PA (2005) De novo identification of repeat families in large genomes. Bioinformatics 21(S1):i351–i358

Raisonnier A (1991) Duplication of the apolipoprotein C-I gene occurred about forty million years ago. J Mol Evol 32(3):211–219

Ray DA, Xing J, Hedges DJ, Hall MA, Laborde ME, Anders BA, White BR, Stoilova N, Fowlkes JD, Landry KE, Chemnick LG, Ryder OA, Batzer MA (2005a) Alu insertion loci and platyrrhine primate phylogeny. Mol Phyl Evol 35(1):117–126

Ray DA, Hedges DJ, Herke SW, Fowlkes JD, Barnes EW, LaVie DK, Goodwin LM, Densmore LD, Batzer MA (2005b) Chompy: an infestation of MITE-like repetitive elements in the crocodilian genome. Gene 362:1–10

Rinehart TA, Grahn RA, Wichman HA (2004) SINE extinction preceded LINE extinction in sigmodontine rodents: implications for retrotranspositional dynamics and mechanisms. Cytogenet Genome Res 110(1–4):416–425

Rokas A, Holland PW (2000) Rare genomic changes as a tool for phylogenetics. TREE 15:454–459

Romualdi C, Balding D, Nasidze IS, Risch G, Robichaux M, Sherry ST, Stoneking M, Batzer MA, Barbujani G (2002) Patterns of human diversity, within and among continents, inferred from biallelic DNA polymorphisms. Genome Res 12:602–612

Roos C, Schmitz J, Zischler H (2004) Primate jumping genes elucidate strepsirrhine phylogeny. Proc Natl Acad Sci USA 101(29):10650–10654

Roy AM, Carroll ML, Kass DH, Nguyen SV, Salem A, Batzer MA, Deininger PL (1999) Recently integrated human Alu repeats: finding needles in the haystack. Genetica 107:149–161

Roy-Engel AM, Carroll ML, Vogel E, Garber RK, Nguyen SV, Salem AH, Batzer MA, Deininger PL (2001) Alu insertion polymorphisms for the study of human genomic diversity. Genetics 159(1):279–290

Roy-Engel AM, Carroll ML, El-Sawy M, Salem AH, Garber RK, Nguyen SV, Deininger PL, Batzer MA (2002) Non-traditional Alu evolution and primate genomic diversity. J Mol Biol 316(5):1033–1040

Ryan SC, Dugaiczyk A (1989) Newly arisen DNA repeats in primate phylogeny. Proc Natl Acad Sci USA 86:9360–9364

Sakamoto K, Okada N (1985) Rodent type 2 Alu family, rat identifier sequence, rabbit C family, and bovine or goat 73-bp repeat may have evolved from tRNA genes. J Mol Evol 22(2):134–140

Salem AH, Kilroy GE, Watkins WS, Jorde LB, Batzer MA (2003a) Recently integrated Alu elements and human genomic diversity. Mol Biol Evol 20(8):1349–1361

Salem AH, Ray DA, Xing J, Callinan PA, Myers JS, Hedges DJ, Garber RK, Witherspoon DJ, Jorde LB, Batzer MA (2003b) Alu elements and hominid phylogenetics. Proc Natl Acad Sci USA 100(22):12787–12791

Salem AH, Ray DA, Hedges DJ, Jurka J, Batzer MA (2005) Analysis of the human Alu Ye lineage. BMC Evol Biol 5(18):1–9

Sasaki T, Takahashi K, Nikaido M, Miura S, Yasukawa Y, Okada N (2004) First application of the SINE (short interspersed repetitive element) method to infer phylogenetic relationships in reptiles: an example from the turtle superfamily Testudinoidea. Mol Biol Evol 21(4):705–715

Sato A, Takezaki N, Tichy H, Figueroa F, Mayer WE, Klein J (2003) Origin and speciation of haplochromine fishes in East African crater lakes investigated by the analysis of their mtDNA, Mhc genes, and SINEs. Mol Biol Evol 20(9):1448–1462

Sheen FM, Sherry ST, Risch GM, Robichaux M, Nasidze I, Stoneking M, Batzer MA, Swergold GD (2000) Reading between the LINEs: genomic variation induced by LINE-1 retrotransposition. Genome Res 10:1496–1508

Schmidt T (1999) LINEs, SINEs and repetitive DNA: non-LTR retrotransposons in plant genomes. Plant Mol Biol 40:903–910

Schmitz J, Ohme M, Zischler H (2001) SINE insertions in cladistic analyses and the phylogenetic affiliations of Tarsius bancanus to other primates. Genetics 157(2):777–784

Schmitz J, Zischler H (2003) Analysis of retrotransposons in dermopterans uncover a new family of tRNA-derived SINEs and support a monophyletic origin of the order primates. Mol Phyl Evol 28:341–349

Schmitz J, Churakov G, Zischler H, Brosius J (2004) A novel class of mammalian-specific tailless retropseudogenes. Genome Res 14(10A):1911–1915

Schmitz J, Roos C, Zischler H (2005) Primate phylogeny: molecular evidence from retroposons. Cytogenet Genome Res 108(1–3):26–37

Shedlock AM, Milinkovitch MC, Okada N (2000) SINE evolution, missing data, and the origin of whales. Syst Biol 49(4):808–817

Shedlock AM, Okada N (2000) SINE insertions: powerful tools for molecular systematics. Bioessays 22(2):148–160

Shedlock AM, Takahashi K, Okada N (2004) SINEs of speciation: tracking lineages with retroposons. TREE 19(10):545–553

Sherry ST, Harpending HC, Batzer MA, Stoneking M (1997) Alu evolution in human populations: using the coalescent to estimate effective population size. Genetics 147(4):1977–1982

Shimamura M, Yasue H, Ohshima K, Abe H, Kato H, Kishiro T, Goto M, Munechika I, Okada N (1997) Molecular evidence from retroposons that whales form a clade within even-toed ungulates. Nature 388(6643):666–670

Pecon-Slattery J, Wilkerson AJP, Murphy WJ, O'Brien SJ (2000) Phylogenetic assessment of introns and SINEs within the Y chromosome using the cat family Felidae as a species tree. Mol Biol Evol 21(12):2299–2309

Springer MS, Stanhope MJ, Madsen O, de Jong WW (2004) Molecules consolidate the placental mammal tree. TREE 19(8):430–438

Stoneking M, Fontius JJ, Clifford SL, Soodyall H, Arcot SS, Saha N, Jenkins T, Tahir MA, Deininger PL, Batzer MA (1997) Alu insertion polymorphisms and human evolution: evidence for a larger population size in Africa. Genome Res 7(11):1061–1071

Szmulewicz MN, Andino LM, Reategui EP, Woolley-Barker T, Jolly CJ, Disotell TR, Herrera RJ (1999) An Alu insertion polymorphism in a baboon hybrid zone. Am J Phys Anthropology 109(1):1–8

Takahashi K, Terai Y, Nishida M, Okada N (1998) A novel family of short interspersed repetitive elements (SINEs) from cichlids: the patterns of insertion of SINEs at orthologous loci support the proposed monophyly of four major groups of cichlid fishes in Lake Tanganyika. Mol Biol Evol 15(4):391–407

Takahashi K, Terai Y, Nishida M, Okada N (2001a) Phylogenetic relationships and ancient incomplete lineage sorting among cichlid fishes in Lake Tanganyika as revealed by analysis of the insertion of retroposons. Mol Biol Evol 18(11):2057–2966

Takahashi K, Nishida M, Yuma M, Okada N (2001b) Retroposition of the AFC family of SINEs (short interspersed repetitive elements) before and during the adaptive radiation of cichlid fishes in Lake Malawi and related inferences about phylogeny. J Mol Evol 53(4–5):496–507

Tatout C, Lavie L, Deragon JM (1998) Similar target site selection occurs in integration of plant and mammalian retroposons. J Mol Evol 47(4):463–470

Tatout C, Warwick S, Lenoir A, Deragon JM (1999) Sine insertions as clade markers for wild crucifer species. Mol Biol Evol 16(11):1614–1621

Terai Y, Takahashi K, Nishida M, Sato T, Okada N (2003) Using SINEs to probe ancient explosive speciation: hidden radiation of African cichlids? Mol Biol Evol 20(6):924–930

Terai Y, Takezaki N, Mayer WE, Tichy H, Takahata N, Klein J, Okada N (2004) Phylogenetic relationships among East African haplochromine fish as revealed by short interspersed elements (SINEs). J Mol Evol 58(1):64–78

Tishkoff SA, Pakstis AJ, Ruano G, Kidd KK (2000) The accuracy of statistical methods for estimation of haplotype frequencies: an example from the CD4 locus. Am J Human Genet 67(2):518–522

Tu Z (1999) Genomic and evolutionary analysis of Feilai, a diverse family of highly reiterated SINEs in the yellow fever mosquito, Aedes aegypti. Mol Biol Evol 16(6):760–772

Ullu E, Tschudi C (1984) Alu sequences are processed 7SL RNA genes. Nature 312:171–172

van de Lagemaat LN, Gagnier L, Medstrand P, Mager DL (2005) Genomic deletions and precise removal of transposable elements mediated by short identical DNA segments in primates. Genome Res 15(9):1243–1249

van Oppen MJH, Rico C, Turner GF, Hewitt GM (2000) Extensive homoplasy, nonstepwise mutations, and shared ancestral polymorphism at a complex microsatellite locus in Lake Malawi cichlids. Mol Biol Evol 17:489–498

Vincent BJ, Myers JS, Ho HJ, Kilroy GE, Walker JA, Watkins WS, Jorde LB, Batzer MA (2003) Following the LINEs: an analysis of primate genomic variation at human-specific LINE-1 insertion sites. Mol Biol Evol 20(8):1338–1348

Vishwanathan H, Edwin D, Usharani MV, Majumder PP (2003) Insertion/deletion polymorphisms in tribal populations of southern India and their possible evolutionary implications. Human Biol 75(6):873–887

Volfovsky N, Haas BJ, Salzberg SL (2001) A clustering method for repeat analysis in DNA sequences. Genome Biol 2(8):RESEARCH0027

Waddell PJ, Kishino H, Ota R (2001) A phylogenetic foundation for comparative mammalian genomics. Genome Informatics 12:141–154

Wang JL, Song MK, Gonder S, Azrak D, Ray A, Batzer MA, Tishkoff SA, Liang P (2006) Whole genome computational comparative genomics: a fruitful approach for ascertaining Alu insertion polymorphisms. Gene 365:11–20

Watkins WS, Ricker CE, Bamshad MJ, Carroll ML, Nguyen SV, Batzer MA, Harpending C, Rogers AR, Jorde LB (2001) Patterns of ancestral human diversity: an analysis of Alu insertion and restriction site polymorphisms. Am J Human Genet 68:738–752

Watkins WS, Rogers AR, Ostler CT, Wooding S, Bamshad MJ, Brassington AM, Carroll ML, Nguyen SV, Walker JA, Prasad BV, Reddy PG, Das PK, Batzer MA, Jorde LB (2003) Genetic variation among world populations: inferences from 100 Alu insertion polymorphisms. Genome Res 13(7):1607–1618

Weiner AM (1980) An abundant cytoplasmic 7S RNA is complementary to the dominant interspersed middle repetitive DNA sequence family in the human genome. Cell 22:209–218

Weiner AM, Deininger PL, Efstratiadis A (1986) Nonviral retroposons: genes, pseudogenes, and transposable elements generated by the reverse flow of genetic information. Annu Rev Biochem 55:631–661

Wulff K, Gazda H, Schroder W, Robicka-Milewska R, Herrmann FH (2000) Identification of a novel large F9 gene mutationan insertion of an Alu repeated DNA element in exon e of the factor 9 gene. Hum Mutat 15:299

Xing J, Salem AH, Hedges DJ, Kilroy GE, Watkins WS, Schienman JE, Stewart CB, Jurka J, Jorde LB, Batzer MA (2003) Comprehensive analysis of two Alu Yd subfamilies. J Mol Evol 57:S76–S89

Xing J, Wang H, Han K, Ray DA, Huang CH, Chemnick LG, Stewart CB, Disotell TR, Ryder OA, Batzer MA (2005) A mobile element based phylogeny of Old World monkeys. Mol Phyl Evol 37(3):872–880

Yang S, Smit AF, Schwartz S, Chiaromonte F, Roskin KM, Haussler D, Miller W, Hardison RC (2004) Patterns of insertions and their covariation with substitutions in the rat, mouse, and human genomes. Genome Res 14:517–527

Genome Dyn Stab (4)
D.-H. Lankenau, J.-N. Volff: Transposons and the Dynamic Genome
DOI 10.1007/7050_2008_041/Published online: 18 October 2008
© Springer-Verlag Berlin Heidelberg 2008

Genome Defense Against Transposable Elements and the Origins of Regulatory RNA

I. King Jordan[1] · Wolfgang J. Miller[2] (✉)

[1]School of Biology, Georgia Institute of Technology, 310 Ferst Drive, Atlanta, GA 30306, USA

[2]Laboratories of Genome Dynamics, Center of Anatomy and Cell Biology, Medical University of Vienna, Währingerstraße 10, 1090 Vienna, Austria
wolfgang.miller@meduniwien.ac.at

Abstract Under selective pressure to contain the harmful effects of transposition, genomes have evolved multiple RNA-based mechanisms for regulating transposable elements (TEs). In this chapter, we describe a number of examples of RNA-based TE defense mechanisms. Once established, these RNA-mediated TE silencing mechanisms, such as RNA interference by miRNAs, may come to be used to regulate host genes. It is becoming possible to reconstruct evolutionary transitions demonstrating how specific TE defense mechanisms were co-opted to provide additional regulatory complexity for host genes. For instance, we have recently shown how miRNAs may have evolved from siRNA encoding TEs. Here we propose another specific model, the transcript infection model, whereby TE insertion dynamics can couple RNA-mediated repression mechanisms to the regulation of host genes.

Abbreviations

dsRNA	Double-stranded RNA
miRNA	MicroRNA
MITE	Miniature inverted repeat transposable element
piRNA	Piwi-interfering RNA
PTGS	Post-transcriptional gene silencing
rasiRNA	Repeat-associated small interfering RNA
RISC	RNA-induced silencing complex
RNAi	RNA interference
siRNA	Small interfering RNA
TAS	Telomeric associated sequence
TE	Transposable element
TIR	Terminal inverted repeat
UTR	Untranslated leader region

1
The Ascent of Regulatory RNA

Two competing epistemologies aim to explain the origins of fundamental changes in scientific thought. Karl Popper was a champion of the principle of falsification, whereby any observation inconsistent with a theory necessitates immediate rejection of that theory (Popper 1959). For Popper, falsifiability distinguished science from non-science. Thomas Kuhn, on the other hand, held that established theories only give way under the weight of an accumulation of multiple anomalies that cannot be explained through the accepted world-view (Kuhn 1962). When this occurs, a paradigm shift ensues resulting in both a new theory and a radically altered world-view. The realization that RNA molecules play a fundamental role in regulating eukaryotic gene expression represents such a paradigm shift in biology (Britten and Davidson 1969). In fact, it took no fewer than three independent discoveries, each in a different phylum, before the broader implications of RNA-mediated gene silencing were fully appreciated and biologists were finally able to accommodate RNA as a regulatory agent alongside protein factors of the classic cis-trans gene regulation model.

RNA-mediated gene silencing was first discovered when botanists working with transgenic plants began to notice a number of confusing gene silencing phenomena (reviewed in Matzke and Matzke 2004). What was then called "co-suppression" was found to occur when plant transgenes involved in pigment synthesis were over-expressed, resulting in silencing of both the transgenes and the homologous plant genes (Napoli, Lemieux and Jorgensen 1990; van der Krol, Mur, Beld et al. 1990). Cosuppression was later determined to occur post-transcriptionally, and while the mechanism underlying these gene silencing events remained unclear, it seemed to be related to sequence interactions that occurred between transgenes with similar (or identical) sequences and/or between transgenes and related plant genes (de Carvalho, Gheysen, Kushnir et al. 1992; van Blokland, van der Geest, Mol et al. 1994). Models of what became known as post-transcriptional gene silencing (PTGS) in plants first articulated the connection between gene silencing and 1-RNA-dependent RNA polymerase, 2-small RNA species and 3-double-stranded RNA (dsRNA) (Lindbo, Silva-Rosales, Proebsting et al. 1993). A few years later, the connection between plant PTGS and sequence-specific RNA-mediated endonucleolytic degradation of mRNA was more definitively established (Metzlaff, O'Dell, Cluster et al. 1997).

In the mean time, a similar transgene induced gene silencing phenomenon, dubbed "quelling", was observed for the fungus *Neurospora crassa* (Romano and Macino 1992). As was seen for plants, exogenous *Neurospora* pigment genes were found to be silenced by homologous transgenes, and the effect appeared to be copy-number dependent. Ultimately, a quelling-defective mutant phenotype (Cogoni and Macino 1997) was shown to be linked to an RNA-dependent RNA polymerase gene (Cogoni and Macino 1999), confirming more speculative models as to the mechanism of RNA-mediated gene silencing.

Finally, observations of complementary antisense RNA sequences pointed to a possible regulatory role for RNA in animals (Lankenau, Corces and

Lankenau 1994), and later RNA-mediated gene silencing was definitively proven in the flatworm *Caenorhabditis elegans*, where it was called RNA interference (RNAi) (Fire, Xu, Montgomery et al. 1998). A major contribution of this work was the demonstration of the role of dsRNA in sequence-specific gene silencing. Fire et al. also showed that a tiny amount of dsRNA was sufficient to cause RNAi, pointing to the likely involvement of amplification in the process. So while the plant and fungal efforts came first, the subsequent *C. elegans* work both unified the understanding of PTGS, quelling and RNAi, and much like Kuhn may have predicted, served as the tipping point after which the fundamental understanding of how eukaryotic genes are regulated was forever changed.

2
RNAi and Genome Defense

Early work on PTGS in plants was also critical in establishing the connection between RNA-mediated gene silencing and genome defense against invading genetic elements. In fact, not long after PTGS was discovered, it became clear that viral sequences could both initiate, and be subject to, PTGS. This was revealed when tobacco plants expressing a viral coat protein encoding transgene recovered from infection with the virus and then became resistant to subsequent viral infection (Lindbo, Silva-Rosales, Proebsting et al. 1993). In addition to virus transgene stimulation of PTGS, plants were also shown to use PTGS as a defense against naturally occurring viral infections (Covey, Al-Kaff, Lángara et al. 1997; Ratcliff, Harrison and Baulcombe 1997). Together, these data led to the notion that a number of gene silencing mechanisms may have originally evolved as genome defense mechanisms that could guard against the harmful effects of viral infection and/or transposition (Matzke, Mette and Matzke 2000). Once these global regulatory mechanisms were established, they could have been co-opted by the host to add an additional layer of regulatory complexity for its own genes. For instance, if an invading genetic element integrated in the flanking region of a host gene, it could change the expression of the adjacent gene as the element sequence was acted on by various host repression mechanisms (Matzke, Mette and Matzke 2000). In addition, the machinery that evolved to defend against invading genetic elements could just as easily be used to regulate host genes, as we believe to be the case for microRNAs (Piriyapongsa, Marino-Ramirez and Jordan 2007). We elaborate on these scenarios and make specific predictions on the insertion patterns and regulatory effects of mobile genetic elements in the section outlining our "transcript infection" model.

Shortly after the role of dsRNA in RNAi was uncovered in *C. elegans*, a connection was made between RNA-mediated gene silencing and transposable elements (TEs). In *C. elegans*, RNAi was initially related to repression of TEs when RNAi deficient mutants were shown to lose the ability to repress Tc1 elements in the germline (Ketting, Haverkamp, van Luenen et al. 1999; Tabara, Sarkissian, Kelly et al. 1999). The specific mechanism underlying Tc1 silencing by RNAi was related to the presence of dsRNAs that are formed when terminal inverted repeat (TIR) sequences at the ends of the elements base pair with each other (Sijen and Plasterk 2003). This base pairing occurs when full-length Tc1 elements are expressed as RNA; since the TIR sequences at the ends of the element are complementary to each other in the single stranded RNA molecule, they can fold into "snap-back" structures forming dsRNA (Fig. 1). The bound TIR dsRNA sequences are cleaved by the RNAi endonucleases to yield short interfering RNAs (siRNAs) that silence element expression via sequence-specific degradation of complementary Tc1 mRNAs, to which they bind. siRNAs encoded by the *MuDR* family of maize elements were also shown to repress TE activity of the same family of elements via specific targeted mRNA degradation (Slotkin, Freeling and Lisch 2005). Thus, TE sequences are both the initiators and the targets of RNAi by siRNAs. Considering these TE-siRNA data, together with the plant findings on PTGS and viral resistance, RNAi was designated as "the genome's immune system" (Plasterk 2002).

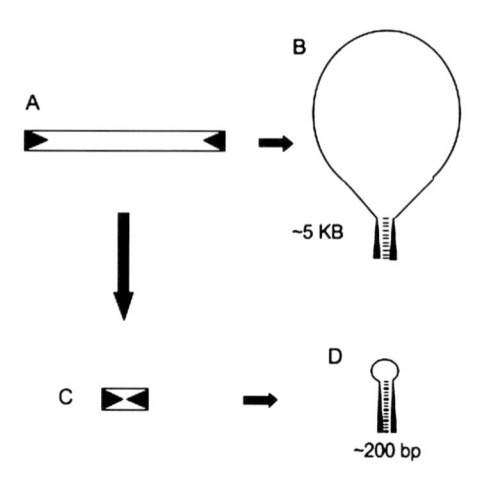

Fig. 1 Model for the siRNA-to-miRNA evolutionary transition. **A** Full-length (autonomous) DNA-type TEs encode siRNAs from snap-back dsRNA regions. (**B**) Formed by the bound terminal inverted repeats (TIRs). **C** Miniature inverted-repeat transposable elements (MITEs) are non-autonomous deletion derivatives of full-length DNA-type TEs, which contain TIRs and short internal regions. **D** MITEs expressed as RNA fold to form hairpin structures that resemble pre-miRNAs. A number of human and plant miRNA genes a derived from MITEs

3
TEs and microRNAs

While the relationship between TEs and RNA-mediated silencing via siRNAs has been appreciated since RNAi was first studied, the TE origins of a related class of regulatory RNAs, microRNAs (miRNAs), were only recently uncovered. miRNAs are also short ($\sim$ 22–25 bp) RNA molecules with a functional role in RNAi analogous to that of siRNAs (Ambros 2004; Bartel 2004). Single stranded mature miRNA sequences with regulatory activity are processed from longer transcripts by two rounds of RNA cleavage. In animals, a relatively long pri-mRNA is cleaved near the base of a hairpin (stem-loop) structure by the enzyme Drosha to yield a $\sim$ 70–90 bp pre-miRNA hairpin. The pre-miRNA hairpin is exported from the nucleus to the cytoplasm and a mature miRNA sequence is then cleaved from the dsRNA stem region by the endonuclease Dicer. Dicer is the same endonuclease that processes siRNAs from longer dsRNA sequences. In plants, both of the mature miRNA biogenesis steps are catalyzed by the related endonuclease Dicer-like1 in the nucleus. Mature miRNAs associate with the RNA-induced silencing complex (RISC), and together the miRNA–RISC targets mRNAs for regulation. miRNA target specificity is determined by partial complementarity with the $3'$-untranslated region (UTR) sequence of the mRNA, and regulation is achieved by translational repression and/or mRNA degradation. miRNAs have been implicated in a variety of functions, including developmental timing (Lee, Feinbaum and Ambros 1993; Reinhart, Slack, Basson et al. 2000), apoptosis (Brennecke, Hipfner, Stark et al. 2003), and hematopoetic differentiation (Chen, Li, Lodish et al. 2004).

The relationship between TEs and miRNAs was discovered when a number of miRNA genes were shown to be derived from TE sequences (Mette, van der Winden, Matzke et al. 2002; Smalheiser and Torvik 2005; Borchert, Lanier and Davidson 2006; Piriyapongsa and Jordan 2007; Piriyapongsa, Marino-Ramirez and Jordan 2007). In the human genome, a group of related miRNA genes was found to be derived from the Made1 family of TEs (Piriyapongsa and Jordan 2007). Made1 elements (Morgan 1995; Oosumi, Belknap and Garlick 1995; Smit and Riggs 1996) are members of a specific class on DNA-type TEs known as miniature inverted-repeat transposable elements (MITEs) (Bureau and Wessler 1992; Bureau and Wessler 1994). MITEs are short non-autonomous derivatives of full-length DNA-type elements (Feschotte and Mouches 2000; Feschotte, Zhang and Wessler 2002). Full-length DNA-type elements are typically several kb in length and contain a single open reading frame, which encodes the transposase enzyme that catalyzes transposition, flanked by two TIR sequences on either end of the elements (Fig. 1A). As is the case with the Tc1 elements of *C. elegans* (Sijen and Plasterk 2003), full-length transcripts of DNA-type elements can fold into "snap-back" structures with the two TIRs forming a dsRNA region (Fig. 1B). This dsRNA region can be processed to yield siRNAs that silence expression of the elements. MITEs are shorter sequences of $\sim$ 80–500 bp, which lack the internal ORF of full-length elements but retain the TIRs (Fig. 1C). In other words, MITEs are closer to being palindromes, and read through transcription of MITEs will lead to RNA sequences that can fold into hairpin structures reminiscent of the pre-miRNA the sequences processed by Dicer to yield mature miRNAs (Fig. 1D).

The difference in structures that produce siRNAs versus miRNAs may be related to the different lengths of the transcripts that are processed including both the bound dsRNA regions and the intervening regions. For instance, longer dsRNA regions are most likely to yield multiple siRNA sequences, while shorter dsRNA hairpins, with small loops, will yield a single miRNA sequence. This may have to do with steric hindrance, i.e., physical-spatial constraints, placed on the dsRNA endonucleolytic machinery that cleaves small hairpins. In other words, RNA endonucleases can scan along longer dsRNA structures to yield multiple siRNAs, but they may only be able to cleave a single miRNA from a short hairpin sequence. It is also worth noting that TE-related dsRNA could be derived from sense and antisense transcripts generated from convergent transcription of the same element and/or read-through transcription of dispersed elements from both directions. Such dsRNA sequences would be expected to yield siRNAs.

The relationship between full-length DNA-type elements and siRNAs on the one hand, and MITEs and miRNAs on the other, led to the articulation of a specific model for how miRNAs could have evolved from siRNA encoding TEs in a stepwise manner (Piriyapongsa and Jordan 2007). As illustrated in Fig. 1, this model posits that siRNAs were first processed from the two TIRs of full-length elements bound as dsRNA. Later, as derivative MITEs evolved from full-length elements and proliferated in the genome, the same RNA endonucleolytic processing machinery cleaved the dsRNA from the hairpin stem regions yielding mature miRNA sequences. A corollary prediction of the TE-miRNA origins model holds that evolutionary intermediates may exist as TE sequences that encode both siRNAs and miRNAs. This prediction was tested and confirmed by an analysis of *Arabidopsis thaliana* and *Oryza sativa* (rice) genomic sequence and expression data (Piriyapongsa and Jordan 2008).

In Arabidopsis and rice, there are in fact a number of examples of individual TE insertions that encode both siRNAs and miRNAs. Dual siRNA-miRNA encoding TEs can be expressed as read-through transcripts from the intronic regions of spliced RNA messages. These TE-transcripts can fold to form the hairpin (stem-loop) structures characteristic of miRNA genes along with longer dsRNA regions that are typically processed as siRNAs. Taken together with a recent study showing Drosha independent processing of miRNAs from Drosophila introns (Ruby, Jan and Bartel 2007), and phylogenetic analysis indicating that Dicer is more ancient than Drosha (Cerutti and Casas-Mollano 2006), these results indicate that ancestral miRNAs could have evolved from TEs prior to the full elaboration of the miRNA biogenesis pathway. Later, as the specific miRNA biogenesis pathway evolved, and numerous other expressed inverted repeat regions came to be recognized by the miRNA processing endonucleases, the host gene related regulatory functions of miRNAs emerged. In this way, host genomes were afforded an additional level of regulatory complexity as a by-product of TE defense mechanisms. The siRNA-to-miRNA evolutionary transition is representative of a number of other RNA-based regulatory mechanisms that initially evolved to silence TEs and were later co-opted to serve as regulators of host gene expression (Girard and Hannon 2008).

4
Repeat-Associated Sequences and piRNAs

In the past few years alone, fundamental TE-research from a number of different labs has significantly expanded our understanding of transposon silencing mechanisms and host genome evolution. Thanks to the completion of the *Drosophila melanogaster* heterochromatin sequencing project, which provides deep insights into the structure and organization of heterochromatin (Smith, Shu, Mungall et al. 2007), the availability of inexpensive mass sequencing technologies such as 454 pyrosequencing, and novel bioinformatics tools, the TE community has revolutionized their concepts on TE biology (recently reviewed in: Aravin, Hannon and Brennecke 2007; Hartig, Tomari and Forstemann 2007; O'Donnell and Boeke 2007).

Whereas in earlier models the actual molecular key-players for TE-silencing were enigmatic, the current TE-silencing model is based on members of Argonaute proteins that play a pivotal role in TE silencing (Aravin, Hannon and Brennecke 2007). These proteins belong to the so-called Argonaute superfamily, bind distinct classes of small RNAs and form the core of the RNA-induced silencing complex (RISC), which is the RNA-interference effector complex (Tolia and Joshua-Tor 2007). Argonaute proteins segregate into two functionally and evolutionarily distinct clades, the Ago clade and the Piwi clade. Whereas in fission yeast and plants only Ago clade proteins are found, ciliates and slime molds encode exclusively Piwi clade proteins. Animal genomes typically contain both clades, and hence, with the further functional dissection of Argonaute proteins, it is becoming clear that the phylogenetic division of Argonautes reflects their underlying biology.

The Ago clade proteins complex with siRNAs and miRNAs, which both derive from dsRNA precursors (Tolia and Joshua-Tor 2007). Whereas miRNA-Ago complexes interfere with the translation and stability of protein-coding mRNAs, by which fine-tuning of gene expression is accomplished, siRNA-Ago complexes are targeted against exogenous viral parasites in animal systems (Wang, Aliyari, Li et al. 2006). In *C. elegans* however, some endogenous siRNAs were found that are likely to participate in host gene regulation as well (Ruby, Jan, Player et al. 2006).

The Piwi clade proteins are found in all animals examined so far but their expression is tightly restricted to the germline. Whereas the three Piwi proteins Aubergine, Piwi, and AGO3 are all expressed in both Drosophila male and female germline (Cox, Chao, Baker et al. 1998; Saito, Nishida, Mori et al. 2006; Brennecke, Aravin, Stark et al. 2007; Gunawardane, Saito, Nishida et al. 2007), expression of the mammalian Piwi homologues MIWI (PIWIL1), MILI (PIWIL2), and MIWI2 (PIWIL4) is restricted to the mouse testes (Kuramochi-Miyagawa, Kimura, Yomogida et al. 2001; Deng and Lin 2002; Kuramochi-Miyagawa, Kimura, Ijiri et al. 2004; Carmell, Girard, van de Kant et al. 2007). Consequently, Piwi loss-of-function mutants generally exhibit massive defects in germ cell development of male and female Drosophila (Cox, Chao, Baker et al. 1998) and during mouse spermatogenesis (Deng and Lin 2002; Kuramochi-Miyagawa, Kimura, Ijiri et al. 2004; Carmell, Girard, van de Kant et al. 2007).

In addition to their fundamental functions in germline development in animals, Piwi proteins play a pivotal role in post-transcriptional silencing of genomic parasites such as TEs (Aravin, Naumova, Tulin et al. 2001; Savitsky, Kwon, Georgiev et al. 2006; Vagin, Sigova, Li et al. 2006). In Piwi loss-of-function mutants of Drosophila, massive bursts of TE transcripts were observed (Aravin, Naumova, Tulin et al. 2001). Later, a novel class of small 25- to 27-nucleotide RNAs with homology to repetitive elements, called repeat-associated small interfering RNAs (rasiRNAs) was discovered (Aravin, Lagos-Quintana, Yalcin et al. 2003). Direct experimental evidence for the conserved functional interaction between Piwi proteins and rasiRNA came from immunoprecipitations of Piwi complexes in Drosophila (Saito, Nishida, Mori et al. 2006; Vagin, Sigova, Li et al. 2006; Brennecke, Aravin, Stark et al. 2007; Gunawardane, Saito, Nishida et al. 2007), Zebrafish (Houwing, Kamminga, Berezikov et al. 2007) and mice (Kim 2006). Due to their tight association with Piwi proteins, and their evolutionary conservation throughout the animal kingdom, rasiRNAs were renamed to Piwi-interfering RNAs (piRNAs).

The Piwi piRNA system forms a separate post-transcriptional regulatory network distinct from the canonical RNAi and miRNA pathways, since Piwis have never been observed to complex with miRNAs in any organism (Aravin, Gaidatzis, Pfeffer et al. 2006; Girard, Sachidanandam, Hannon et al. 2006; Saito, Nishida, Mori et al. 2006; Brennecke, Aravin, Stark et al. 2007). Similar to miRNAs, the piRNAs carry a $5'$-monophosphate group and exhibit a preference for a $5'$-uridine residue (Aravin, Gaidatzis, Pfeffer et al. 2006; Girard, Sachidanandam, Hannon et al. 2006; Lau, Seto, Kim et al. 2006). Unlike animal miRNAs, but similar to plant miRNAs, piRNAs carry a $2'$-O-methyl modification at their $3'$ ends added by a Hen-1 family RNA methyltransferase (Hartig, Tomari and Forstemann 2007). The Piwi-piRNA pathway is largely independent from the function of Dicer that is the key-player of miRNA and siRNA pathways and is clearly distinctive in terms of evolutionary age and mode of selection.

Whereas some miRNAs are conserved over millions of years by coevolving in concert with their host-derived target genes under purifying selection, piRNAs evolve rapidly and even closely related species harbor different repertoires of piRNAs cocktails (Aravin, Gaidatzis, Pfeffer et al. 2006; Girard, Sachidanandam, Hannon et al. 2006; Lau, Seto, Kim et al. 2006). These data suggest that piRNAs are under rapid adaptive evolution in concert with horizontally invading and vertically expanding genomic parasites (Aravin, Hannon and Brennecke 2007). Consequently, Piwi proteins that are tightly linked to dynamic piRNAs should also evolve rapidly in an adaptive manner similar to the situation found in centromere-specific histone variants and their hyperdynamic satellite DNA targets (Henikoff, Ahmad and Malik 2001; Malik, Vermaak and Henikoff 2002) or the antiviral DNA-editing enzyme APOBEC3G and their viral targets (Sawyer, Emerman and Malik 2004).

With the isolation and characterization of Piwi-associated piRNAs by immunoprecipitation and large scale sequencing, their genomic mapping analysis revealed a limited set of discrete loci that could give rise to most piRNAs (Brennecke, Aravin, Stark et al. 2007; Gunawardane, Saito, Nishida et al. 2007). These so-called piRNA clusters map to the pericentromeric or telomeric heterochromatin, range from several to hundreds of kilobases, are devoid of protein-coding genes, and are densely occupied by eroded remnants of formerly active TEs and other repeats. Similar piRNA clusters were found in mammals (Aravin, Gaidatzis, Pfeffer et al. 2006; Girard, Sachidanandam, Hannon et al. 2006; Lau, Seto, Kim et al. 2006) and zebrafish (Houwing, Kamminga, Berezikov et al. 2007).

As one of the most extensively studied piRNA clusters, the flamenco locus, serves as a model system for expanding our understanding on transposon silencing. Thanks to elegant genetic studies, this locus has been mapped to the X-chromosomal pericentromeric heterochromatin of Drosophila as a key-regulator for controlling transposon activity (Pelisson, Song, Prud'homme et al. 1994; Prud'homme, Gans, Masson et al. 1995). Mutations in this master locus give rise to transcriptional derepression and transpositional bursts of formerly silenced retrotransposons, such as gypsy, Idefix and ZAM (Pelisson, Song, Prud'homme et al. 1994; Bucheton 1995; Prud'homme, Gans, Masson et al. 1995; Desset, Meignin, Dastugue et al. 2003; Meignin, Dastugue and Vaury 2004). Subsequently, Pelisson and colleagues demonstrated that flamenco-mediated silencing of gypsy is dependent on the presence of the Piwi protein (Sarot, Payen-Groschene, Bucheton et al. 2004). As deduced from recent immunoprecipitations and large-scale sequence analyses (Brennecke, Aravin, Stark et al. 2007), the main fraction of Drosophila piRNAs were bound by and isolated from the Piwi protein complex in the germline maps almost exclusively to the minus strand of the flamenco locus (Brennecke, Aravin, Stark et al. 2007; Gunawardane, Saito, Nishida et al. 2007). This piRNA master locus spans a region of 179-kb and consists of eighty-seven percent of nested mobile DNAs of many TE families. Importantly the three retrotransposons ZAM, Idefix and gypsy were all found in the minus strand orientation within the flamenco locus (Brennecke, Aravin, Stark et al. 2007). This remarkable strand asymmetry suggests that the heterochromatic flamenco locus is expressed via follicle cells in the germline as a long primary transcript composed of numerous TE-remnants, all in antisense orientation.

In the course of two independent piRNA studies, the authors detected a remarkable strand bias of piRNAs derived from each of the three Piwi complexes (Brennecke, Aravin, Stark et al. 2007; Gunawardane, Saito, Nishida et al. 2007). Whereas Piwi and Aubergine proteins preferentially bind piRNAs corresponding to the antisense strand of transposons, Ago3 complexes almost exclusively with piRNAs of TEs in sense orientation. Surprisingly, a unique complementary relationship between sense and antisense piRNAs was observed over 10 nucleotides in the 5′ ends between Aubergine- or Piwi- and Ago3-associated piRNAs. Based on these observations, both research groups concluded that Piwi-mediated cleavage events generate new piRNAs in a self-reinforcing amplification cycle for piRNA generation named the "Ping-Pong" model (Fig. 2) (Brennecke, Aravin, Stark et al. 2007; Gunawardane, Saito, Nishida et al. 2007). According to this model, initiation of the cycle begins with processing of primary piRNAs, transcribed by defective TE-remnants in the heterochromatin. These early piRNAs are antisense to active, euchromatic TE copies and bind either Piwi or Aubergine. Triggered by partial complementarity in their 5′ ends, the Aubergine/Piwi-piRNA complexes target and cleave their active euchromatic transposon-counterparts and generate new sense piRNAs that bind Ago3. Next, the sense piRNA-Ago3 complex directs another cleavage event of a heterochromatic piRNA cluster transcript creating a new antisense piRNA capable of binding to Piwi and Aubergine. This model allows the efficient amplification of weak piRNA signals in the absence of a RNA-dependent RNA polymerase, which is an essential enzyme for RNAi signal-reinforcement in fission yeast, plants and nematodes (Cogoni and Macino 1999; Dalmay, Hamilton, Rudd et al. 2000; Sijen and Kooter 2000), but is absent in Drosophila and vertebrates (Schwarz, Hutvagner, Haley et al. 2002). Clearly, piRNAs provide an excitingly dynamic system for acquiring adaptive immunity against invading genomic parasites.

A second Drosophila piRNA cluster that has been genetically linked to TE control corresponds to the telomeric associated sequence (TAS) repeat on the X-chromosome (X-TAS). Insertions of one or two P elements into X-TAS are sufficient for silencing in trans P element mobility in the P-M hybrid dysgenesis system (Ronsseray, Lehmann and Anxolabehere 1991; Marin, Lehmann, Nouaud et al. 2000; Stuart, Haley, Swedzinski et al. 2002) and to reactivate P-transposition in mutants of the Piwi family (Reiss, Josse, Anxolabehere et al. 2004; Josse, Teysset, Todeschini et al. 2007). The exact insertion positions were mapped and sequenced within X-TAS repeats (Karpen

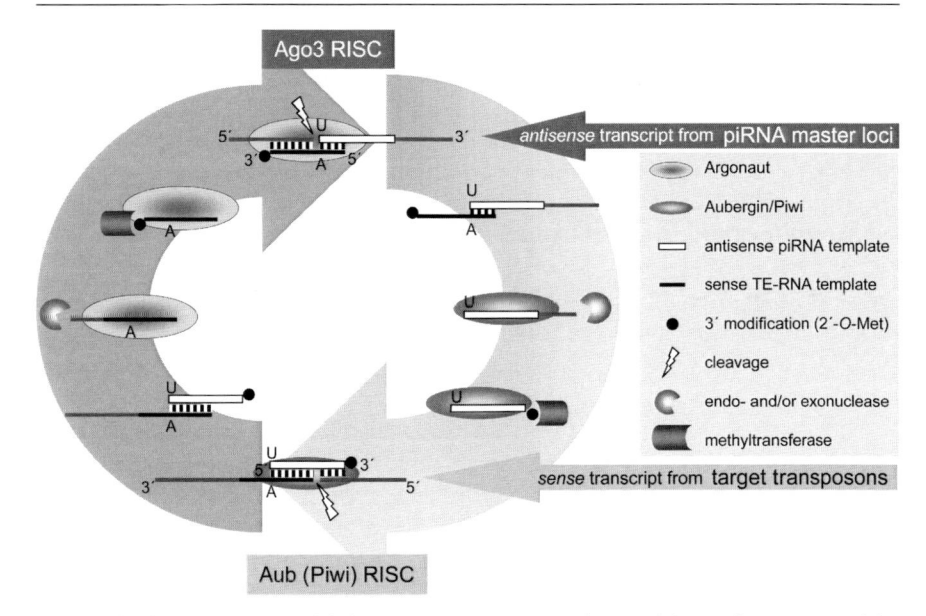

Fig. 2 The "Ping-Pong" model for piRNA biogenesis. The model was first proposed by Brennecke et al. 2007 and Gunawardane et al. 2007 and the figure is modified after Hartig, Tomari and Förstemann 2007. *Bottom*: Transposon-derived sense transcripts expressed from multiple genomic sites are targeted and cleaved by Piwi or Aub RISC loaded with a piRNA guide. The cleaved transcript is only partly degraded and serves as a specified "hunting-bait" for Ago3 RISC. *Top*: This complex in turn cleaves the antisense transcripts that originate from the master control loci. Again, the cleaved RNA serves as a bait for Piwi or Aub RISC. Thus, sense and antisense transcripts fuel an amplification cycle in which the 5′ ends of piRNAs are defined by RISC cleavage. In this model, the 3′ ends are shortened by an endonuclease and/or exonuclease, subsequently 2′-O-Me-modified by a methyltransferase

and Spradling 1992; Ronsseray, Lehmann, Nouaud et al. 1997; Boivin, Gally, Netter et al. 2003) and multiple small RNAs homologous to X-TAS were isolated by (Brennecke, Aravin, Stark et al. 2007). Unfortunately, in the cases of both Drosophila strains, the strain that was analyzed for piRNA expression (Brennecke, Aravin, Stark et al. 2007; Gunawardane, Saito, Nishida et al. 2007) and the strain sequenced for the Drosophila heterochromatin sequencing project (Hoskins, Carlson, Kennedy et al. 2007; Smith, Shu, Mungall et al. 2007) are devoid of P elements, since in both cases P element-free lab strains (M strains) were used in the analyses.

Despite our current lack of sequence information on P element-derived piRNA in the germline of Drosophila, it seems very likely that in more recent *D. melanogaster* strains harboring P elements in the subtelomeric cluster the X-TAS locus is actually driving the expression of such integrated P elements in an antisense orientation. Similar to the Ping-Pong model (Brennecke, Aravin, Stark et al. 2007; Gunawardane, Saito, Nishida et al. 2007) determined from the pericentromeric flamenco piRNA cluster, the heterochromatic X-TAS antisense-P-transcripts could be targeted by the Piwi protein complex, which binds euchromatic sense-P-transcripts in a sequence specific manner by complementarity and chops them into short piRNAs that further trigger the sequence-specific degradation of novel P-transcripts.

Since parasitic P elements have invaded natural *D. melanogaster* populations within the last 50–100 years by horizontal transfer from neotropical Drosophila host species (Kidwell 1983), the piRNA mediated resistance locus was born immediately when P elements inserted stochastically into X-TAS of the subtelomere and were trapped. Through this random insertion event, a new TE-specific immune locus was generated, which is now expressing RNA-based "antibodies" against recently invading and selfishly expanding euchromatic P elements (O'Donnell and Boeke 2007), providing an adaptive, vertically transmitted immunity against P-mobility to the following generations. Hence, the classic P element system presents a perfect model system for studying the exceptional dynamics of the interplay between genome parasites and hosts even over a relatively short time scale.

5
Transcript Infection Model

The data described above, emerging from experimental analyses of a number of different RNA-based regulatory systems, all point to TEs as being initiators of, and targeted by, RNA-mediated antisense gene regulation. While the impetus for the evolutionary origins of these distinct regulatory systems was the need to repress TEs, in many cases host genomes have co-opted the regulatory complexity afforded by RNA-based systems to control expression of their own genes. Here, we propose a specific model that links the RNA-mediated repression of TE sequences to the regulation of host genes via "transcript infection."

Transcript infection occurs when a TE inserts inside of, or adjacent to, a host gene (Fig. 3). The abundance of TEs, coupled with their transpositional activity, ensures that many gene-associated insertions of this kind will occur. Such insertion events may allow for the production of chimeric transcripts that include both host gene and TE sequences. Most studies to date have emphasized the functional consequences of TE insertions inside genes or in upstream proximal promoter regions. Our transcript infection model rests on TE insertions that occur downstream (3′) of host genes. These downstream insertions may lead to substantial regulatory effects by inducing antisense regulation of nearby host genes.

In Fig. 3, we illustrate a number of specific scenarios by which transcript infection via 3′ TE insertions could lead to the repression of host genes. For instance, a 3′ TE insertion oriented antisense to the host gene could give

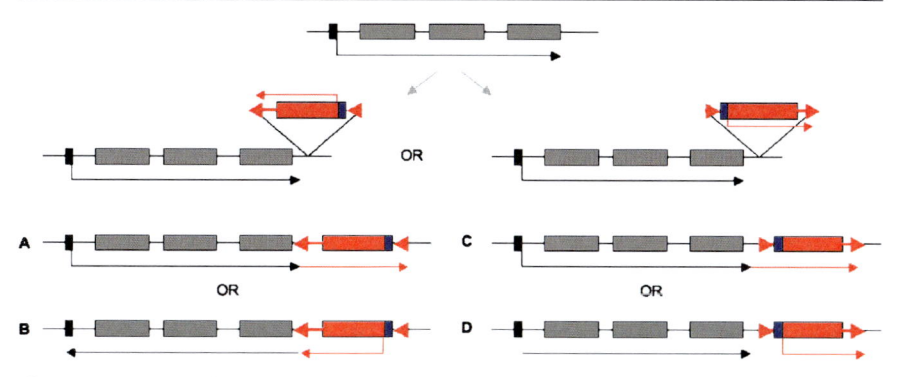

Fig. 3 Transcript infection model. TEs insert downstream of host genes in the antisense or sense orientation. Antisense oriented downstream TEs can result in host gene sense-TE antisense chimeric transcripts promoted by the host gene (**A**) or TE sense-host gene antisense chimeric transcripts promoted by the TE (**B**). Sense oriented downstream TEs can result in host gene sense-TE sense chimeric transcripts promoted by the host gene (**C**) or independent transcripts promoted by the host gene and the TE (**D**). The implications of these distinct scenarios for RNA-mediated host gene regulation are detailed in the text

rise to a chimeric transcript, promoted by the host gene, consisting of sense host gene transcript fused to an antisense TE transcript (Fig. 3A). In this case, Piwi and Aubergine could complex with the antisense TE sequence and titrate off the host transcript, leading to down regulation or loss of function. This would be analogous to a TE-induced auto-immune disease. On the other hand, if transcription is promoted by the antisense oriented 3′ TE insertion, a read-through transcript consisting of both sense TE and antisense host gene sequence could be produced (Fig. 3B). This chimeric transcript could be targeted for miRNA interference mediated by Argonaute and Dicer proteins or silencing via the piRNA pathway. In either case, there would be a reduced level of host transcript dependent upon the transcriptional activity of the TE promoter.

Sense oriented 3′ TE insertions would lead to different predictions regarding RNA-mediated regulation of host genes. For instance, the host gene promoter could drive transcription of chimeric mRNAs that include both sense host gene and sense TE sequences (Fig. 3C). The sense TE sequences could be targeted by the piRNA pathway via Ago3. This would cause depletion of the fused host gene transcript similar to what was proposed for the antisense oriented 3′ insertions. However, if there is independent transcription of the host gene and the 3′ sense TE insertion, then one would neither expect any transcriptional interference, nor, consequently, an effect on host gene expression (Fig. 3D).

There are a few caveats with respect to the details of the transcription infection model that we would like to describe. For instance, in order for TE transcription infection events to exert regulatory effects, antisense TE transcripts incorporated into a sense primary gene transcript would need to be complexed by Piwi and Aubergine in the piRNA pathway. It is possible that only some of the antisense TE transcripts produced in this will feed into the piRNA pathway. There may be some determining factor that influences such an outcome such as the stage of chimeric expression and/or the temporal and spatial availability of triggering cellular factors. A deeper understanding of such factors will have to await further experimental proof. In addition, sense TE transcripts fused with primary gene transcripts will only be incorporated into the piRNA or miRNA regulatory pathways if piRNAs or miRNAs against that particular element already exist. Finally, there are substantial differences in the number (density) of TE sequences in different species and these differences may well relate to the probability of the transcription infection model to exert substantial regulatory effects.

Consistent with the transcript infection model, we recently showed that human TEs at the 3′ ends of genes are prone to produce antisense transcription and that these TE antisense promoters are evolutionarily conserved, suggesting some regulatory function (Conley, Miller and Jordan 2008). Given the availability of complete genome sequences, along with TE annotations and high-throughput transcript mapping, it should be possible to more directly test the predictions of the transcript infection model.

References

Ambros V (2004) The functions of animal microRNAs. Nature 431:350–355
Aravin A, Gaidatzis D, Pfeffer S, Lagos -Quintana M, Landgraf P, Iovino N et al. (2006) A novel class of small RNAs bind to MILI protein in mouse testes. Nature 442:203–207
Aravin AA, Hannon GJ, Brennecke J (2007) The Piwi-piRNA pathway provides an adaptive defense in the transposon arms race. Science 318:761–764
Aravin AA, Lagos-Quintana M, Yalcin A, Zavolan M, Marks D, Snyder B et al. (2003) The small RNA profile during Drosophila melanogaster development. Dev Cell 5:337–350
Aravin AA, Naumova NM, Tulin AV, Vagin VV, Rozovsky YM, Gvozdev VA (2001) Double-stranded RNA-mediated silencing of genomic tandem repeats and transposable elements in the D. melanogaster germline. Curr Biol 11:1017–1027
Bartel DP (2004) MicroRNAs: genomics, biogenesis, mechanism, and function. Cell 116:281–297
Boivin A, Gally C, Netter S, Anxolabehere D, Ronsseray S (2003) Telomeric associated sequences of Drosophila recruit polycomb-group proteins in vivo and can induce pairing-sensitive repression. Genetics 164:195–208
Borchert GM, Lanier W, Davidson BL (2006) RNA polymerase III transcribes human microRNAs. Nat Struct Mol Biol 13:1097–1101
Brennecke J, Aravin AA, Stark A, Dus M, Kellis M, Sachidanandam R et al. (2007) Discrete small RNA-generating loci as master regulators of transposon activity in Drosophila. Cell 128:1089–1103
Brennecke J, Hipfner DR, Stark A, Russell RB, Cohen SM (2003) Bantam encodes a developmentally regulated microRNA that controls cell proliferation and regulates the proapoptotic gene hid in Drosophila. Cell 113:25–36

Britten RJ, Davidson EH (1969) Gene regulation for higher cells: a theory. Science 165:349–357

Bucheton A (1995) The relationship between the flamenco gene and gypsy in *Drosophila*: how to tame a retrovirus. Trends Genet 11:349–353

Bureau TE, Wessler SR (1992) Tourist: a large family of small inverted repeat elements frequently associated with maize genes. Plant Cell 4:1283–1294

Bureau TE, Wessler SR (1994) Stowaway: a new family of inverted repeat elements associated with the genes of both monocotyledonous and dicotyledonous plants. Plant Cell 6:907–916

Carmell MA, Girard A, Van de Kant HJ, Bourc'his D, Bestor TH, De Rooij DG et al. (2007) MIWI2 is essential for spermatogenesis and repression of transposons in the mouse male germline. Dev Cell 12:503–514

Cerutti H, Casas-Mollano JA (2006) On the origin and functions of RNA-mediated silencing: from protists to man. Curr Genet 50:81–99

Chen CZ, Li L, Lodish HF, Bartel DP (2004) MicroRNAs modulate hematopoietic lineage differentiation. Science 303:83–86

Cogoni C, Macino G (1997) Isolation of quelling-defective (qde) mutants impaired in posttranscriptional transgene-induced gene silencing in *Neurospora crassa*. Proc Natl Acad Sci USA 94:10233–10238

Cogoni C, Macino G (1999) Gene silencing in *Neurospora crassa* requires a protein homologous to RNA-dependent RNA polymerase. Nature 399:166–169

Conley AB, Miller WJ, Jordan IK (2008) Human *cis* natural antisense transcripts initiated by transposable elements. Trends Genet 24:53–56

Covey SN, Al-Kaff NS, Lángara A, Turner DS (1997) Plants combat infection by gene silencing. Nature 385:781–782

Cox DN, Chao A, Baker J, Chang L, Qiao D, Lin H (1998) A novel class of evolutionarily conserved genes defined by piwi are essential for stem cell self-renewal. Genes Dev 12:3715–3727

Dalmay T, Hamilton A, Rudd S, Angell S, Baulcombe DC (2000) An RNA-dependent RNA polymerase gene in *Arabidopsis* is required for posttranscriptional gene silencing mediated by a transgene but not by a virus. Cell 101:543–553

De Carvalho F, Gheysen G, Kushnir S, Van Montagu M, Inze D, Castresana C (1992) Suppression of beta-1,3-glucanase transgene expression in homozygous plants. EMBO J 11:2595–2602

Deng W, Lin H (2002) Miwi, a murine homolog of piwi, encodes a cytoplasmic protein essential for spermatogenesis. Dev Cell 2:819–830

Desset S, Meignin C, Dastugue B, Vaury C (2003) COM, a heterochromatic locus governing the control of independent endogenous retroviruses from *Drosophila melanogaster*. Genetics 164:501–509

Feschotte C, Mouches C (2000) Evidence that a family of miniature inverted-repeat transposable elements (MITEs) from the *Arabidopsis thaliana* genome has arisen from a pogo-like DNA transposon. Mol Biol Evol 17:730–737

Feschotte C, Zhang X, Wessler SR (2002) Miniature inverted-repeat transposable elements and their relationships to established DNA transposons. In: Craig N, Craigie R, Gellert M, Lambowitz A (eds) Mobile DNA II. ASM Press, Washington, DC, pp 1147–1158

Fire A, Xu S, Montgomery MK, Kostas SA, Driver SE, Mello CC (1998) Potent and specific genetic interference by double-stranded RNA in *Caenorhabditis elegans*. Nature 391:806–811

Girard A, Hannon GJ (2008) Conserved themes in small-RNA-mediated transposon control. Trends Cell Biol 18:136–148

Girard A, Sachidanandam R, Hannon GJ, Carmell MA (2006) A germline-specific class of small RNAs binds mammalian Piwi proteins. Nature 442:199–202

Gunawardane LS, Saito K, Nishida KM, Miyoshi K, Kawamura Y, Nagami T et al. (2007) A slicer-mediated mechanism for repeat-associated siRNA 5′ end formation in Drosophila. Science 315:1587–1590

Hartig JV, Tomari Y, Forstemann K (2007) piRNAs – the ancient hunters of genome invaders. Genes Dev 21:1707–1713

Henikoff S, Ahmad K, Malik HS (2001) The centromere paradox: stable inheritance with rapidly evolving DNA. Science 293:1098–1102

Hoskins RA, Carlson JW, Kennedy C, Acevedo D, Evans-Holm M, Frise E et al. (2007) Sequence finishing and mapping of Drosophila melanogaster heterochromatin. Science 316:1625–1628

Houwing S, Kamminga LM, Berezikov E, Cronembold D, Girard A, Van den Elst H et al. (2007) A role for Piwi and piRNAs in germ cell maintenance and transposon silencing in Zebrafish. Cell 129:69–82

Josse T, Teysset L, Todeschini AL, Sidor CM, Anxolabehere D, Ronsseray S (2007) Telomeric trans-silencing: an epigenetic repression combining RNA silencing and heterochromatin formation. PLoS Genet 3:1633–1643

Karpen GH, Spradling AC (1992) Analysis of subtelomeric heterochromatin in the Drosophila minichromosome Dp1187 by single P element insertional mutagenesis. Genetics 132:737–753

Ketting RF, Haverkamp TH, Van Luenen HG, Plasterk RH (1999) Mut-7 of C. elegans, required for transposon silencing and RNA interference, is a homolog of Werner syndrome helicase and RNaseD. Cell 99:133–141

Kidwell MG (1983) Evolution of hybrid dysgenesis determinants in Drosophila melanogaster. Proc Natl Acad Sci USA 80:1655–1659

Kim VN (2006) Small RNAs just got bigger: Piwi-interacting RNAs (piRNAs) in mammalian testes. Genes Dev 20:1993–1997

Kuhn TS (1962) The Structure of Scientific Revolutions. University of Chicago Press, Chicago

Kuramochi-Miyagawa S, Kimura T, Ijiri TW, Isobe T, Asada N, Fujita Y et al. (2004) Mili, a mammalian member of piwi family gene, is essential for spermatogenesis. Development 131:839–849

Kuramochi-Miyagawa S, Kimura T, Yomogida K, Kuroiwa A, Tadokoro Y, Fujita Y et al. (2001) Two mouse piwi-related genes: miwi and mili. Mech Dev 108:121–133

Lankenau S, Corces VG, Lankenau D-H (1994) The Drosophila micropia retrotransposon encodes a testis-specific antisense RNA complementary to reverse transcriptase. Mol Cell Biol 14:1764–1775

Lau NC, Seto AG, Kim J, Kuramochi-Miyagawa S, Nakano T, Bartel DP et al. (2006) Characterization of the piRNA complex from rat testes. Science 313:363–367

Lee RC, Feinbaum RL, Ambros V (1993) The C. elegans heterochronic gene lin-4 encodes small RNAs with antisense complementarity to lin-14. Cell 75:843–854

Lindbo JA, Silva-Rosales L, Proebsting WM, Dougherty WG (1993) Induction of a Highly Specific Antiviral State in Transgenic Plants: Implications for Regulation of Gene Expression and Virus Resistance. Plant Cell 5:1749–1759

Malik HS, Vermaak D, Henikoff S (2002) Recurrent evolution of DNA-binding motifs in the Drosophila centromeric histone. Proc Natl Acad Sci USA 99:1449–1454

Marin L, Lehmann M, Nouaud D, Izaabel H, Anxolabehere D, Ronsseray S (2000) P-Element repression in Drosophila melanogaster by a naturally occurring defective telomeric P copy. Genetics 155:1841–1854

Matzke MA, Matzke AJ (2004) Planting the seeds of a new paradigm. PLoS Biol 2:E133

Matzke MA, Mette MF, Matzke AJ (2000) Transgene silencing by the host genome defense: implications for the evolution of epigenetic control mechanisms in plants and vertebrates. Plant Mol Biol 43:401–415

Meignin C, Dastugue B, Vaury C (2004) Intercellular communication between germ line and somatic line is utilized to control the transcription of ZAM, an endogenous retrovirus from *Drosophila melanogaster*. Nucleic Acids Res 32:3799–3806

Mette MF, Van der Winden J, Matzke M, Matzke AJ (2002) Short RNAs can identify new candidate transposable element families in *Arabidopsis*. Plant Physiol 130:6–9

Metzlaff M, O'Dell M, Cluster PD, Flavell RB (1997) RNA-mediated RNA degradation and chalcone synthase A silencing in petunia. Cell 88:845–854

Morgan GT (1995) Identification in the human genome of mobile elements spread by DNA-mediated transposition. J Mol Biol 254:1–5

Napoli C, Lemieux C, Jorgensen R (1990) Introduction of a chimeric chalcone synthase gene into Petunia results in reversible co-suppression of homologous genes in trans. Plant Cell 2:279–289

O'Donnell KA, Boeke JD (2007) Mighty Piwis defend the germline against genome intruders. Cell 129:37–44

Oosumi T, Belknap WR, Garlick B (1995) Mariner transposons in humans. Nature 378:672

Pelisson A, Song SU, Prud'homme N, Smith PA, Bucheton A, Corces VG (1994) Gypsy transposition correlates with the production of a retroviral envelope-like protein under the tissue-specific control of the *Drosophila* flamenco gene. EMBO J 13:4401–4411

Piriyapongsa J, Jordan IK (2007) A Family of Human MicroRNA Genes from Miniature Inverted-Repeat Transposable Elements. PLoS ONE 2:e203

Piriyapongsa J, Jordan IK (2008) Dual coding of siRNAs and miRNAs by plant transposable elements. RNA 14:814–821

Piriyapongsa J, Marino-Ramirez L, Jordan IK (2007) Origin and evolution of human microRNAs from transposable elements. Genetics 176:1323–1337

Plasterk RH (2002) RNA silencing: the genome's immune system. Science 296:1263–1265

Popper KR (1959) The Logic of Scientific Discovery. Hutchinson, London

Prud'homme N, Gans M, Masson M, Terzian C, Bucheton A (1995) Flamenco, a gene controlling the gypsy retrovirus of *Drosophila melanogaster*. Genetics 139:697–711

Ratcliff F, Harrison BD, Baulcombe DC (1997) A similarity between viral defense and gene silencing in plants. Science 276:1558–1560

Reinhart BJ, Slack FJ, Basson M, Pasquinelli AE, Bettinger JC, Rougvie AE et al. (2000) The 21-nucleotide let-7 RNA regulates developmental timing in *Caenorhabditis elegans*. Nature 403:901–906

Reiss D, Josse T, Anxolabehere D, Ronsseray S (2004) Aubergine mutations in *Drosophila melanogaster* impair P cytotype determination by telomeric P elements inserted in heterochromatin. Mol Genet Genomics 272:336–343

Romano N, Macino G (1992) Quelling: transient inactivation of gene expression in *Neurospora crassa* by transformation with homologous sequences. Mol Microbiol 6:3343–3353

Ronsseray S, Lehmann M, Anxolabehere D (1991) The maternally inherited regulation of P elements in *Drosophila melanogaster* can be elicited by two P copies at cytological site 1A on the X chromosome. Genetics 129:501–512

Ronsseray S, Lehmann M, Nouaud D, Anxolabehere D (1997) P element regulation and X-chromosome subtelomeric heterochromatin in *Drosophila melanogaster*. Genetica 100:95–107

Ruby JG, Jan C, Player C, Axtell MJ, Lee W, Nusbaum C et al. (2006) Large-scale sequencing reveals 21U-RNAs and additional microRNAs and endogenous siRNAs in *C. elegans*. Cell 127:1193–1207

Ruby JG, Jan CH, Bartel DP (2007) Intronic microRNA precursors that bypass Drosha processing. Nature 448:83–86

Saito K, Nishida KM, Mori T, Kawamura Y, Miyoshi K, Nagami T et al. (2006) Specific association of Piwi with rasiRNAs derived from retrotransposon and heterochromatic regions in the *Drosophila* genome. Genes Dev 20:2214–2222

Sarot E, Payen-Groschene P, Bucheton A, Pelisson A (2004) Evidence for a piwi-dependent RNA silencing of the gypsy endogenous retrovirus by the *Drosophila melanogaster* flamenco gene. Genetics 166:1313–1321

Savitsky M, Kwon D, Georgiev P, Kalmykova A, Gvozdev V (2006) Telomere elongation is under the control of the RNAi-based mechanism in the *Drosophila* germline. Genes Dev 20:345–354

Sawyer SL, Emerman M, Malik HS (2004) Ancient adaptive evolution of the primate antiviral DNA-editing enzyme APOBEC3G. PLoS Biol 2:E275

Schwarz DS, Hutvagner G, Haley B, Zamore PD (2002) Evidence that siRNAs function as guides, not primers, in the *Drosophila* and human RNAi pathways. Mol Cell 10:537–548

Sijen T, Kooter JM (2000) Post-transcriptional gene-silencing: RNAs on the attack or on the defense? Bioessays 22:520–531

Sijen T, Plasterk RH (2003) Transposon silencing in the *Caenorhabditis elegans* germ line by natural RNAi. Nature 426:310–314

Slotkin RK, Freeling M, Lisch D (2005) Heritable transposon silencing initiated by a naturally occurring transposon inverted duplication. Nat Genet 37:641–644

Smalheiser NR, Torvik VI (2005) Mammalian microRNAs derived from genomic repeats. Trends Genet 21:322–326

Smit AF, Riggs AD (1996) Tiggers and DNA transposon fossils in the human genome. Proc Natl Acad Sci USA 93:1443–1448

Smith CD, Shu S, Mungall CJ, Karpen GH (2007) The Release 5.1 annotation of *Drosophila melanogaster* heterochromatin. Science 316:1586–1591

Stuart JR, Haley KJ, Swedzinski D, Lockner S, Kocian PE, Merriman PJ et al. (2002) Telomeric P elements associated with cytotype regulation of the P transposon family in *Drosophila melanogaster*. Genetics 162:1641–1654

Tabara H, Sarkissian M, Kelly WG, Fleenor J, Grishok A, Timmons L et al. (1999) The rde-1 gene, RNA interference, and transposon silencing in *C. elegans*. Cell 99:123–132

Tolia NH, Joshua-Tor L (2007) Slicer and the argonautes. Nat Chem Biol 3:36–43

Vagin VV, Sigova A, Li C, Seitz H, Gvozdev V, Zamore PD (2006) A distinct small RNA pathway silences selfish genetic elements in the germline. Science 313:320–324

Van Blokland R, Van der Geest N, Mol JNM, Kooter JM (1994) Transgene-mediated suppression of chalcone synthase expression in Petunia hybrida results from an increase in RNA turnover. Plant J 6:861–877

Van der Krol AR, Mur LA, Beld M, Mol JN, Stuitje AR (1990) Flavonoid genes in petunia: addition of a limited number of gene copies may lead to a suppression of gene expression. Plant Cell 2:291–299

Wang XH, Aliyari R, Li WX, Li HW, Kim K, Carthew R et al. (2006) RNA interference directs innate immunity against viruses in adult *Drosophila*. Science 312:452–454

Genome Dyn Stab (4)
D.-H. Lankenau, J.-N. Volff: Transposons and the Dynamic Genome
DOI 10.1007/7050_2009_045/Published online: 26 May 2009
© Springer-Verlag Berlin Heidelberg 2009

When *Drosophila* Meets Retrovirology: The *gypsy* Case

Christophe Terzian[1] (✉) · Alain Pelisson[2] · Alain Bucheton[2]

[1]INRA, EPHE, ENVL, UMR754 "Rétrovirus et Pathologie Comparée", Université de Lyon, Université Lyon 1, 50 avenue Tony Garnier, 69007 Lyon, France
cterzian@univ-lyon1.fr

[2]Institut de Génétique Humaine, Centre National de la Recherche Scientifique, 34396 Montpellier Cedex 5, France

Abstract Insect endogenous retroviruses (IERVs) are present in the genome of several species. It was shown that *gypsy* is an active endogenous retrovirus in *Drosophila melanogaster*, which could be transmitted to individuals fed with *gypsy*-containing extracts. Moreover, *gypsy* replication depends on cell-cell transfer. Here, we review recent findings which help to elucidate the structure and the role of *gypsy* Env protein during *gypsy* horizontal and vertical transfers.

Abbreviations

ERVs   Endogenous retroviruses
IERV   Insect endogenous retrovirus
LTR   Long terminal repeat
piRNA   Piwi-interacting RNA
Env   Envelope glycoprotein
EM   Electron microscopy

1
Introduction

Since the discovery of viruses, their origins remain a puzzling challenge not yet resolved, even if attractive scenarios have been recently proposed (Forterre 2006; Koonin et al. 2008). Unfortunately for virologists, no ancestral viral forms are detected within fossil organisms in order to infer the evolutionary history of viruses. However, a family of viruses remains suitable for evolutionary studies: the retroviridae. Indeed, the retroviruses are the only known viruses which are able to infect and colonize their host's germline genomes, i.e. endogenous retroviruses (see Box 1), giving the opportunity to trace back the time of their integration using the phylogenetic relationships of their hosts. Endogenous retroviruses (ERVs) are present in a large spectrum of organisms, from plants to animals. One particularity of ERVs is their close structural similarity to LTR-retrotransposons, which led Temin to propose

Box 1

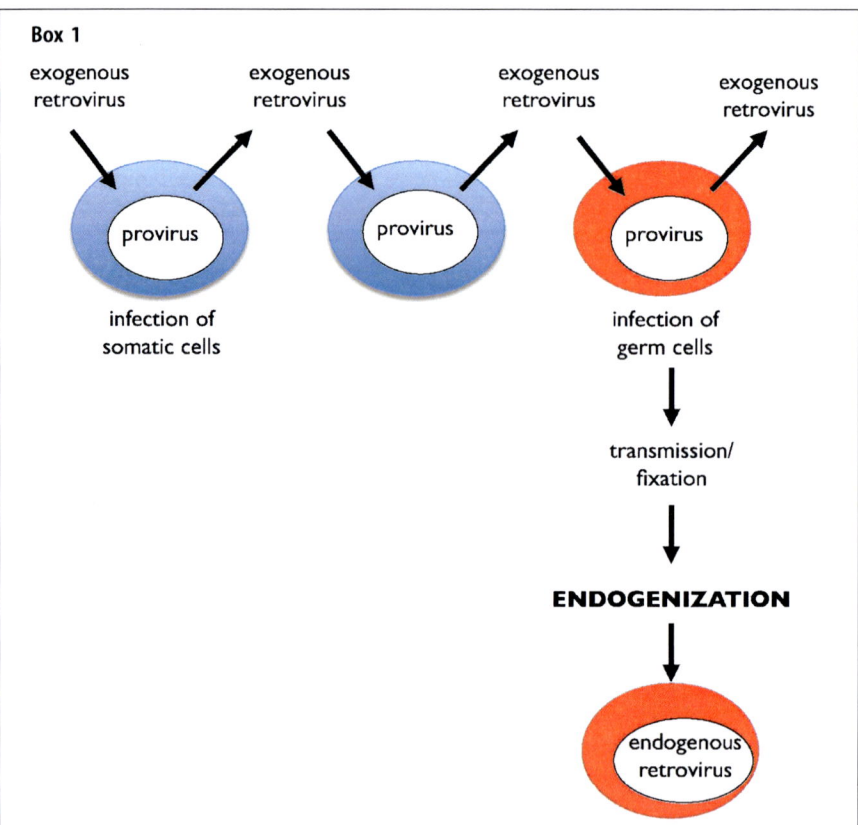

Exogenous retroviruses (such as HIV-1) infect somatic cells , but some can also infect germline cells (cells that make eggs and sperm) and once they have done so and have been transmitted to the next generation, they are termed endogenous. Endogenous retroviruses are then transmitted through the germline as stable Mendelian genes. Endogenous retroviruses can persist in the genome of their host for long periods. They are generally transcriptionally silent and are often defective, differing from the exogenous counterpart by deletions or point mutations that render them incapable of forming infectious virus. However, several ERVs maintain at least some intact open reading frames that can be expressed. Particularly, it has been shown that several ERVs containing *gag* and *pol* genes also retain the ability to encode functional Env glycoproteins. This is the case for the human HERV-K and HERV-Fc, porcine ERVs, ovine enJSRV and insect ERVs. One can ask why some ERV genomes have kept the ability to encode a functional envelope. For several ERVs, ERV ORFs have been characterized either as playing a physiological role (i.e. "domestication", see I.K. Jordan and W.J. Miller, this BOOK), or as protecting the host from exogenous counterparts. A third explanation concerns the fitness of the ERVs themselves, and a special focus on the role of the Gag and Env glycoprotein on the maintenance of ERVs in their hosts has been set. These three topics (domestication, restriction, ERV fitness) concern the fundamental physiological

---

Box 1 (continued)
and developmental processes as well as the mechanisms developed by host
genomes to protect against infection and the emergence of new viruses which could
result from the interplay between endogenous and exogenous retroviruses.

---

the hypothesis that retroviruses evolved from retrotransposons (Temin 1980).
Reciprocally, LTR-retrotransposons could represent the next step of ERVs
evolution, resulting from a balance between selection, loss of viral genes
and effective host population size (Katzourakis et al. 2005). In any case, one
should consider two kinds of ERVs: the inactive ERVs, which have accumulated indels[1] and mutations over time and are currently unable to produce
proteins; and the active ERVs, which are able to produce proteins and to move
either intracellulary from one chromosomal position to another, or from cell
to cell. Among the host organisms, *Drosophila melanogaster* is probably one
of the most powerful models for the study of the genesis and impact of ERVs,
since its genome contains the highest percentage of active full-length endogenous retroviruses among animals studied until now (Eickbush and Furano
2002). The aim of this review is to provide recent insights into the evolution of the *gypsy* retroelement, one of the most studied insect endogenous
retroviruses (IERVs).

2
Historical Background

gypsy, also named *mdg-4*, was described in the 1980s as an active LTR-
retrotransposon, able to induce mutations by insertion into cellular genes
(Modolell et al. 1983; Bayev et al. 1984). However, *gypsy* reached a new status
when its nucleotide sequence pointed out the presence of three open reading frames similar to the Gag, Pol, and Env retroviral proteins (Marlor et al.
1986). At this time, presence of a third ORF was also found in a few *Drosophila*
elements like *297*, *17.6* and *tom* (Inouye et al. 1984,1986; Tanda et al. 1994).
Moreover, it was shown that *tom* and *gypsy* formed spliced subgenomic transcripts able to encode their ORF3 products, in the same way as that used
by retroviruses to express their envelope proteins (Pelisson et al. 1994; Tanda
et al. 1994) (Fig. 1). Such similarities sustained the hypothesis that *gypsy* was
an infectious element due to the properties of its envelope protein. Two independent experiments provided results in favor of the "infection hypothesis":

---

[1] The term "indel" is used in its modern meaning here. It is an artificial blend of the two words
"insertions" and "deletions". Contrasting point mutations indels refer to two types of genetic mutations that are often considered together because of their similar molecular origin, similar effect and
the inability to distinguish between them in computational comparisons of two sequences. Indels
are useful genetic markers in natural populations and in phylogenetic studies.

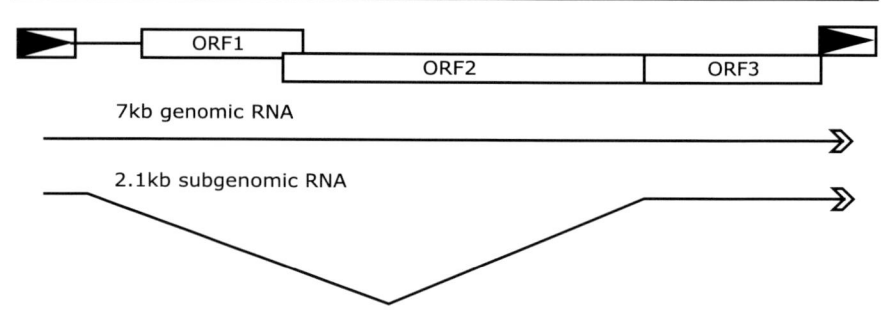

Fig. 1 Structural features of *gypsy*. The organization of *gypsy* is shown, with the three open reading frames (ORFs) corresponding to the *gag*, *pol*, and *env* genes present in retroviruses. These coding regions are flanked by two Long Terminal Repeats (LTRs). The 7 kb RNA is the template for reverse transcription and is used to synthesize the products of ORF1 and ORF2. A splicing event that produces the 2.1 kb RNA generates an AUG codon that can be used to initiate ORF3 translation

a high level of *gypsy* insertion was observed in the progeny of *Drosophila* larvae fed with either (1) extracts derived from smashed pupae (Kim et al. 1994), or (2) sucrose gradient fractions containing *gypsy* particles (Song et al. 1994). One should admit that it was an unusual way to study infectious properties of an element using *Drosophila* genetical readouts. However, no ex vivo system is (yet) available to produce *gypsy* particles and to manipulate them using "classical" virological methods. Thanks to a powerful genetic screen based on the high specificity of *gypsy* insertions into the *ovo* gene (Mével-Ninio et al. 1989), the horizontal transfer of *gypsy* from the external medium to the germline of individuals, using apparently the nutrient pathway, was strongly suggested.

3
Finding the Road to the Germline

Several key points remain to be elucidated in order to understand how the primary infection did occur during the ingestion of *gypsy* particles and thereafter their transfer to the germline. In order to get deeper into the last part of the infection process, i.e. the transport of *gypsy* from soma to oocyte, we have to remember the life cycle of *gypsy* (Fig. 2). It was shown during the past years that *gypsy* RNA and proteins are mainly expressed in follicular cells of nonrestrictive, i.e. permissive females (Pelisson et al. 1994; Lecher et al. 1997; Song et al. 1997). Females are permissive if they are defective for the production of specific piRNAs able to target *gypsy* RNAs (Pelisson et al. 2007; Brennecke et al. 2007). It was shown that these piRNAs are encoded by the *flamenco* locus, at position 20A in the pericentromeric region of the X chromosome (Prud'homme et al. 1995; Brennecke et al. 2007; see Box 2). Then,

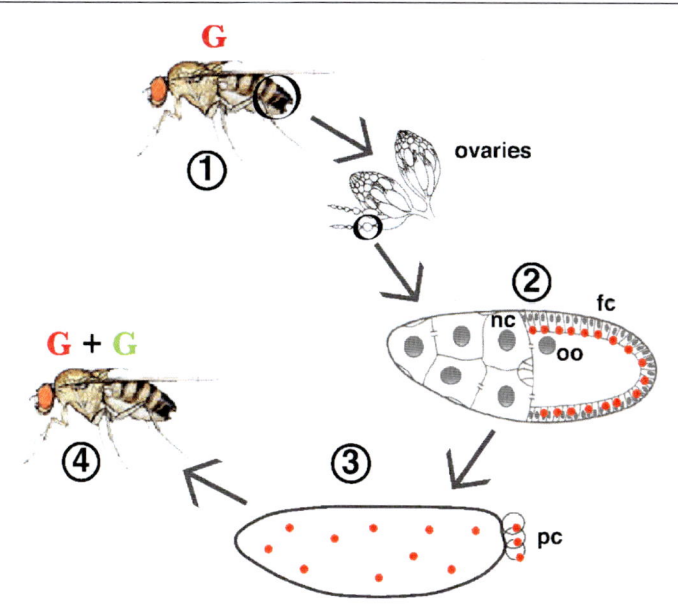

Fig. 2 Gypsy replication cycle. 1. A *flamenco* permissive female containing active *gypsy* proviral copies ("G" in *red*). 2. A stage 9 egg chamber, where *gypsy* particles (*red circles*) are mainly expressed in somatic follicle cells (fc) surrounding the oocyte (oo); nc: nurse cells. 3. The *gypsy* particles are then transferred to the oocyte and the fertilized egg by a yet unknown mechanism; pc–pole cells. 4. The F1 flies will then contain new *gypsy* proviral copies ("G" in *green*) inserted into their somatic and germ nuclei. The expression of *gypsy* is repressed in *flamenco* restrictive females (not shown)

transposition of *gypsy* occurs into the germline of the female lacking *gypsy* piRNAs. Hence, the life cycle of *gypsy* in permissive females recapitulates the soma to germline transfer of *gypsy* which might occur during the last step of the experimental infection described above, i.e. the endogenization process. It was assumed that the *gypsy* Env protein was playing a major role during the replication cycle of *gypsy* (Pelisson et al. 1994). We will present the facts in favor and against this hypothesis.

4
Origin of the *gypsy* Env

Because of the similarity between *gypsy* and retroviral genomic structures, the *gypsy* Env amino acid sequence was first annotated according to the canonical vertebrate Env sequences, i.e. a glycosylated polyprotein with a *N*-terminal signal peptide, cleaved at a furin cleavage motif, giving rise to the SU protein which tethers the virus to the host-cell, and the TM protein which anchors the Env complex to the viron core and mediates the host-cell

Box 2   The increase of *gypsy* copy number is controlled by *flamenco* (*flam*), a large heterochromatic locus in the X chromosome (Bucheton 1995) which could only be sequenced over 180 kb. This locus initially posed a puzzle because it does not contain any single copy sequence that would encode a putative "transposon repressor". It is, instead, mainly made of remnants of dozens of ancestral LTR retroelements, mostly inserted opposite to the transcription orientation of the locus. Since the discovery of 25- to 27-nt antisense *gypsy* RNAs, *flam* and *piwi* have been assumed to be involved in a homology-dependent transposon defense system (Sarot et al. 2004). Mével-Ninio et al. (2007) found that *flam* mutants not only disrupt the *gypsy* repression but also a previously described (Desset et al. 2003) repression of the *ZAM* and *Idefix* LTR elements; hence the hypothesis that the *flam* function is to prevent the expression, in the ovarian somatic cells, of any transposable element family that has at least one complementary sequence in this regulatory locus. Further genetic and molecular evidence supported the notion that the small RNAs, called Piwi-interacting RNAs (piRNAs), are bona fide transposon regulators that are encoded by defective retroelements inserted into *flam* (Pélisson et al. 2007, Brennecke et al. 2007, Mével-Ninio et al. 2007). These piRNAs may originate from a very long precursor RNA molecule, since mutations mapping at the beginning of the *flam* locus disrupt piRNAs some 168 kb away (Brennecke et al. 2007). Expression of the *flam* locus in the ovarian somatic cells is not only required for the proper control of *gypsy*, *ZAM* and *Idefix* but also for the normal development of the somatic and germinal ovarian tissues (Mével-Ninio et al. 2007). Whether some *flam*-encoded piRNAs are responsible for this new function is still unknown.

fusion process (Pelisson et al. 1994). However, most surprisingly, a similarity search in protein databanks identified IERV env-like sequences in several baculovirus genomic sequences (Malik et al. 2000; Rohrmann and Karplus 2001). Baculoviruses represent a family of enveloped viruses with double-stranded DNA genomes (Jehle et al. 2006). They are pathogenic mostly to *Lepidoptera*, but also to *Diptera* and *Hymenoptera*. The IERV env-like sequences present in the baculovirus sequences correspond to viral genes encoding a fusion protein named F protein, which is responsible for tethering the virus to cells and mediating the virus–cell fusion as well as the cell–cell fusion process (Ijkel et al. 2000). It should also be noted that a full-length IERV-like element was previously described in the genome of a baculovirus infecting the *Trichoplasia ni* (Friesen and Nissen 1990). Altogether, these data suggest that *env* gene lateral transfer did occur from a viral genome to another, and baculovirus genomes may represent a landing platform where IERV can insert and capture baculovirus F protein genes (Malik et al. 2000; Rohrmann and Karplus 2001). However, the presence of host insect genes encoding proteins with similarity to baculovirus F proteins in several insect genomes supports the possibility that the baculovirus fusion gene may have been acquired from a cellular gene. The *Drosophila melanogaster* gene named *CG4715* according to the Flybase annotation symbol, was initially described as related to the baculoviral and IERV fusion protein genes (Rohrmann and Karplus 2001). It was then named *Dm-cF* (Lung and Blissard 2005), and finally renamed *iris* which is at present the Flybase official name (Malik and Henikoff 2005). A comprehensive phylogenetic study of *iris* homologues in insect genomes was performed, and the authors suggested that *iris* was domesticated from an IERV more than 25 million years ago, and was then subject to positive selection for a function which is yet unknown (Malik and Henikoff 2005). However, one cannot rule out the opposite hypothesis, i.e. F and IERV Env proteins genes were acquired from an *iris* cellular gene, and then evolved toward acquiring fusion properties. Indeed, *Drosophila* Iris lacks most of the structural properties of a fusion protein, i.e. the furin cleavage site, the fusion peptide and the coiled-coil domain and it only contains a signal peptide and a transmembrane domain (Fig. 3). Moreover, a functional analysis in *Drosophila melanogaster* S2 cells indicated that Iris is localized to the membrane of intracellular organelles, and not to the cellular membrane as *gypsy* Env (Lung and Blissard 2005).

It was clearly demonstrated that F proteins mediate membrane fusion (Westenberg et al. 2004). The significant similarity between IERV Env and baculovirus F sequences may suggest that IERV Env might have also kept fusogenic properties. This hypothesis is now supported by structural and functional analysis as shown below.

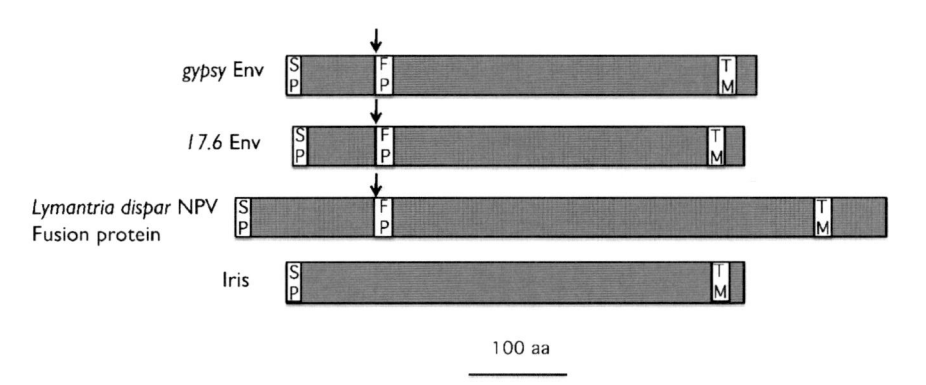

100 aa

Fig. 3 Schematic ORFs of *gypsy* Env, 17.6 Env , the fusion protein of *Lymantria dispar* nucleopolyhedrovirus, and Iris from *D. melanogaster*. SP–predicted signal peptide, FP–the conserved block common to insect retroviruses and baculoviruses fusion protein which includes the fusion peptide and the coiled-coil region, TM–transmembrane region. The *vertical arrows* indicate the furin-like proteolytic cleavage sites that have been experimentally determined. The conserved block FP is characterized by the prosite pattern: R-x(2)-R-x(5,14)-G-x(3)-[KHN]-x(3)-G-x(2)-D-x(2)-D

5
Structural Analysis of *gypsy* Env

The mechanism by which an enveloped virus penetrates a cell is a complex process involving several domains of viral fusion proteins (Weissenhorn et al. 2007). Among them, the role of the fusion peptide is predominant. This hydrophobic sequence is located at the *N*-terminal end of the transmembrane subunit generated by the post-translational cleavage of the class-I envelope protein (Fig. 3). It is involved in the anchoring of the fusion protein into the target membrane. Alignment of several predicted IERV envelope proteins and baculovirus F proteins sequences has led to the identification of a region of high similarity (Misseri et al. 2003). This region could be the fusion peptide because it contains an *RxxR* furin-like cleavage site, contiguous with a hydrophobic sequence, which contains a set of conserved amino acids followed by a predicted coiled-coil region (Fig. 4). The *gypsy* Env sequence contains a RIAR furin-like consensus motif, contiguous with a set of conserved amino acids followed by a predicted coiled-coil region, suggesting that cleavage at

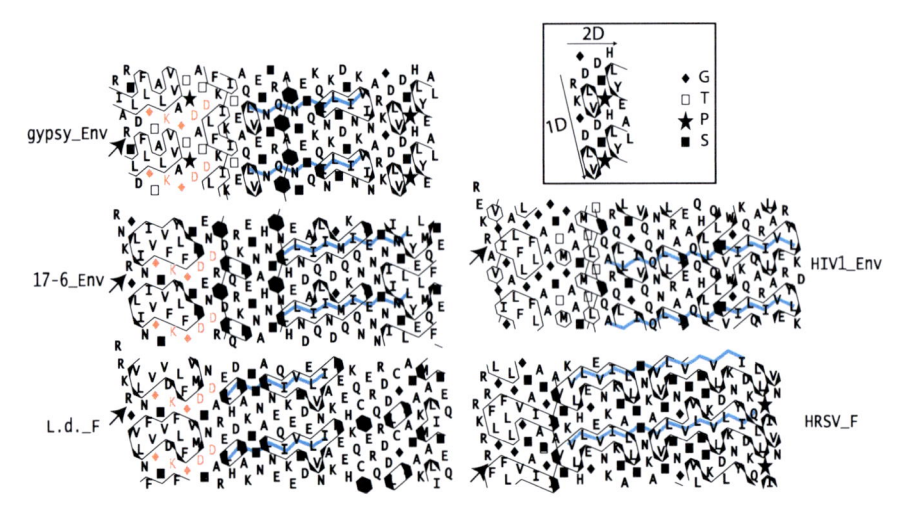

Fig. 4 Hydrophobic cluster analysis (Callebaut et al. 1997) of the *gypsy* Env, 17.6 Env and *Lymantria dispar* NPV Fusion protein (L.d. F) sequences. The amino acid sequences are written vertically (with an angle to the right) on a duplicated alpha-helical net, and the clusters formed by contiguous hydrophobic residues are *boxed*. Four *symbols* are used regarding the specific structural behavior of the amino acids they represent: a *star* for proline, a *diamond* for glycine, a *square* for threonine, and a *square with a dot inside* for serine. The position of the fusion peptides is indicated in *red*, the furin cleavage site is indicated by *arrows*, and the coiled-coil helical structure indicated by the *blue line*. As a comparison, the *N*-term domains of HIV-1 transmembrane and HSRV fusion proteins are shown

RIAR may occur to yield an ectodomain and a transmembrane domain, giving rise to fusion activity by exposing the peptide fusion.

6
Functional Analysis of *gypsy* Env

Although several IERVs are now well described, the functional properties of IERV Env were only addressed for *gypsy*. A first clue toward identifying the infectious properties of *gypsy* Env was obtained when the "feeding" experiment was modified by the addition of anti-Env antibody to the sucrose gradient fractions containing *gypsy* particles: the number of *gypsy* insertions into the progeny genomes was then slightly reduced, suggesting that the mAbs can partially inhibit the horizontal transfer of *gypsy* (Song et al. 1994). A second result in favor of the infectious role of *gypsy* Env was obtained when it was shown that a Moloney leukemia virus vector pseudotyped with *gypsy* Env was able to transduce *Drosophila* cells, albeit with a low efficiency (Teysset et al. 1998). Moreover, *gypsy* Env expressing Sf9 cells are able to fuse with wild-type Sf9 cells (Misseri et al. 2004). These results suggest that *gypsy* Env has fusogenic properties. Because of the structural similarities between the F protein and *gypsy* Env sequences, the role of the RIAR motif as a cleavage site was addressed in vivo and ex vivo by Western blotting experiments (Misseri et al. 2004; Pearson and Rohrmann 2004). The results support the hypothesis that RIAR is cleaved in permissive ovaries as well as in S2 and Sf9 insect culture cells.

7
Role of *gypsy* Env

As described previously, *gypsy* is expressed in permissive follicle cells (somatic cells surrounding the germline) and then integrates into the oocyte. Hence, the *gypsy* replication cycle seems to involve cell-to-cell transfer. The fusogenic properties of *gypsy* Env led us to propose a role for this protein during its transfer to the oocyte. Nonenveloped intracytoplasmic particles containing *gypsy* RNAs were observed by EM in the apical region of permissive follicle cells close to the oocyte, whereas neither particles nor Env were detected in the nonpermissive cells (Lecher et al. 1997). Although those particles were found to accumulate close to *gypsy* Env-containing junction-like membrane domains, enveloped *gypsy* particles budding from the follicle cells were never detected. A few *gypsy* particles were observed within the permissive oocytes: where they came from and how they entered the oocyte remain open questions. In order to address directly the role of Env in *gypsy* replication, an *env*-defective *gypsy* was introduced by P-transgenesis into a permissive strain devoid of active *gypsy* proviruses: it was shown that this *gypsy* construct was still able to transpose, i.e. to be transferred from soma to germline (Chalvet et al. 1999). We cannot rule out that the *env*-defective *gypsy* could be pseudotyped by an unidentified Env-related protein produced by the genome of the transgenic line analyzed. A second explanation could be that the cell-to-cell transfer process is enhanced by the presence of Env but does not absolutely require it. Hence, this result strongly suggests that replication of *gypsy* is not strictly dependent on the *gypsy env* gene. However, if Env is dispensable for *gypsy* replication, what is the mechanism responsible for the soma-to-germline transfer of *gypsy*? This question was addressed with the ZAM element, a closely related IERV from *Drosophila melanogaster*. As for *gypsy*, ZAM particles were seen by EM close to the apical border of the follicle cells (Leblanc et al. 2000). Distribution of ZAM particles was compared between follicle cells bearing different mutations involved in the transport of yolk protein (Brasset et al. 2006). The results suggest that a link exists between the trafficking of ZAM and of vitellogenin from the follicle cells to the oocyte.

Nevertheless, we expect *gypsy* Env has a specific function because selective pressure for the conservation of its function was detected by Ka/Ks tests (Terzian et al. 2000; Llorens et al. 2008). Several studies indicate that *gypsy* sequence phylogeny is incongruent with host phylogeny, which could be explained by several horizontal transfers of *gypsy* between host species (Alberola and de Frutos 1996; Terzian et al. 2000; Vázquez-Manrique et al. 2000; Herédia et al. 2004). Horizontal transfer is a strategy that transposons use to escape mechanisms of repression by the host (Jordan et al. 1999). One reason for the conservation of *gypsy* Env might be that they allow infectious horizontal transfers of *gypsy* among *Drosophila* species without the need of a vector.

8
Conclusion

Exogenous retroviruses were considered to be restricted to vertebrates. This view is now challenged by insect retroviruses like *gypsy*. The *gypsy* element may represent two different steps in the evolutionary history of an endogenous retrovirus: *gypsy* may still reflect the "real-time" process of endogenization of a putative exogenous counterpart, and its regulation by the host genome will help to understand the post-endogenization steps which follow infection of the germline. The *gypsy* Env protein represents a major component of these mechanisms, and understanding its role is a challenge for the next years.

Acknowledgements We thank Dr. Dirk Lankenau and two anonymous reviewers for their helpful comments. We also thank our collaborators in Gif-sur-Yvette and Montpellier. This work was supported by CNRS, EPHE, and FINOVI.

References

Alberola TM, De Frutos R (1996) Molecular structure of a *gypsy* element of *Drosophila subobscura* (*gypsy*Ds) constituting a degenerate form of insect retroviruses. Nucl Acids Res 24:914 923

Bayev AA Jr, Lyubomirskaya NV, Dzhumagaliev EB, Ananiev EV, Amiantova IG, Ilyin YV (1984) Structural organization of transposable element mdg4 from *Drosophila melanogaster* and a nucleotide sequence of its long terminal repeats. Nucl Acids Res 12:3707–3723

Brasset E, Taddei AR, Arnaud F, Faye B, Fausto AM, Mazzini M, Giorgi F, Vaury C (2006) Viral particles of the endogenous retrovirus ZAM from *Drosophila melanogaster* use a pre-existing endosome/exosome pathway for transfer to the oocyte. Retrovirology 3:25

Brennecke J, Aravin AA, Stark A, Dus M, Kellis M, Sachidanandam R, Hannon GJ (2007) Discrete small RNA-generating loci as master regulators of transposon activity in *Drosophila*. Cell 128:1089–1103

Brennecke J, Malone CD, Aravin AA, Sachidanandam R, Stark A, Hannon GJ (2008) An epigenetic role for maternally inherited piRNAs in transposon silencing. Science 322:1387–1392

Bucheton A (1995) The relationship between the *flamenco* gene and gypsy in *Drosophila*: how to tame a retrovirus. Trends Genet 11:349–353

Callebaut I, Labesse G, Durand P, Poupon A, Canard L, Chomilier J, Henrissat B, Mornon JP (1997) Deciphering protein sequence information through hydrophobic cluster analysis (HCA): current status and perspectives. Cell Mol Life Sci 53:621–645

Chalvet F, Teysset L, Terzian C, Prud'homme N, Santamaria P, Bucheton A, Pélisson A (1999) Proviral amplification of the *Gypsy* endogenous retrovirus of *Drosophila melanogaster* involves env-independent invasion of the female germline. EMBO J 18:2659–2669

Desset S, Meignin C, Dastugue B, Vaury C (2003) COM, a heterochromatic locus governing the control of independent endogenous retroviruses from *Drosophila melanogaster*. Genetics 164:501–509

Eickbush TH, Furano AV (2002) Fruit flies and humans respond differently to retrotransposons. Curr Opin Genet Dev 12:669–674

Forterre P (2006) Three RNA cells for ribosomal lineages and three DNA viruses to replicate their genomes: a hypothesis for the origin of cellular domain. Proc Natl Acad Sci USA 103:3669–3674

Friesen PD, Nissen MS (1990) Gene organization and transcription of TED, a lepidopteran retrotransposon integrated within the baculovirus genome. Mol Cell Biol 10:3067–3077

Herédia F, Loreto EL, Valente VL (2004) Complex evolution of *gypsy* in Drosophilid species. Mol Biol Evol 21:1831–1842

IJkel WF, Westenberg M, Goldbach RW, Blissard GW, Vlak JM, Zuidema D (2000) A novel baculovirus envelope fusion protein with a proprotein convertase cleavage site. Virology 275:30–41

Inouye S, Yuki S, Saigo K (1986) Complete nucleotide sequence and genome organization of a *Drosophila* transposable genetic element, 297. Eur J Biochem 154:417–425

Inouye S, Yuki S, Saigo K (1984) Sequence-specific insertion of the *Drosophila* transposable genetic element 17.6. Nature 310:332–333

Jehle JA, Blissard GW, Bonning BC, Cory JS, Herniou EA, Rohrmann GF, Theilmann DA, Thiem SM, Vlak JM (2006) On the classification and nomenclature of baculoviruses: a proposal for revision. Arch Virol 151:1257–1266

Jordan IK, Matyunina LV, McDonald JF (1999) Evidence for the recent horizontal transfer of long terminal repeat retrotransposon. Proc Natl Acad Sci USA 96:12621–12625

Katzourakis A, Rambaut A, Pybus OG (2005) The evolutionary dynamics of endogenous retroviruses. Trends Microbiol 13:463–468

Kim A, Terzian C, Santamaria P, Pélisson A, Purd'homme N, Bucheton A (1994) Retroviruses in invertebrates: the *gypsy* retrotransposon is apparently an infectious retrovirus of *Drosophila melanogaster*. Proc Natl Acad Sci USA 15:1285–1289

Koonin EV, Wolf YI, Nagasaki K, Dolja VV (2008) The Big Bang of picorna-like virus evolution antedates the radiation of eukaryotic supergroups. Nat Rev Microbiol 6:925–939

Leblanc P, Desset S, Giorgi F, Taddei AR, Fausto AM, Mazzini M, Dastugue B, Vaury C (2000) Life cycle of an endogenous retrovirus, ZAM, in *Drosophila melanogaster*. J Virol 74:10658–10669

Llorens JV, Clark JB, Martínez-Garay I, Soriano S, De Frutos R, Martínez-Sebastián MJ (2008) *Gypsy* endogenous retrovirus maintains potential infectivity in several species of Drosophilids. BMC Evol Biol 8:302

Lung O, Blissard GW (2005) A cellular *Drosophila melanogaster* protein with similarity to baculovirus F envelope fusion proteins. J Virol 79:7979–7989

Lécher P, Bucheton A, Pélisson A (1997) Expression of the *Drosophila* retrovirus *gypsy* as ultrastructurally détectable particles in the ovaries of flies carrying a permissive *flamenco* allele. J Gen Virol 78:2379–2388

Malik HS, Henikoff S (2005) Positive selection of Iris, a retroviral envelope-derived host gene in *Drosophila melanogaster*. PLoS Genet 1:e44

Malik HS, Henikoff S, Eickbush TH (2000) Poised for contagion: evolutionary origins of the infectious abilities of invertebrate retroviruses. Genome Res 10:1307–1318

Marlor RL, Parkhurst SM, Corces VG (1986) The *Drosophila melanogaster gypsy* transposable element encodes putative gene products homologous to retroviral proteins. Mol Cell Biol 6:1129–1134

Misseri Y, Cerutti M, Devauchelle G, Bucheton A, Terzian C (2004) Analysis of the *Drosophila gypsy* endogenous retrovirus envelope glycoprotein. J Gen Virol 85:3325–3331

Misseri Y, Labesse G, Bucheton A, Terzian C (2003) Comparative sequence analysis and predictions for the envelope glycoproteins of insect endogenous retroviruses. Trends Microbiol 11:253–256

Modolell J, Bender W, Meselson M (1983) *Drosophila melanogaster* mutations suppressible by the suppressor of Hairy-wing are insertions of a 7.3-kilobase mobile element. Proc Natl Acad Sci USA 80:1678–1682

Mével-Ninio M, Mariol MC, Gans M (1989) Mobilization of the *gypsy* and copia retrotransposons in *Drosophila melanogaster* induces reversion of the ovo dominant female-sterile mutations: molecular analysis of revertant alleles. EMBO J 8:1549–1558

Mével-Ninio M, Pelisson A, Kinder J, Campos AR, Bucheton A (2007) The *flamenco* locus controls the *gypsy* and ZAM retroviruses and is required for *Drosophila* oogenesis. Genetics 175:1615–1624

Pearson MN, Rohrmann GF (2004) Conservation of a proteinase cleavage site between an insect retrovirus (*gypsy*) Env protein and a baculovirus envelope fusion protein. Virology 322:61–68

Prud'homme N, Gans M, Masson M, Terzian C, Bucheton A (1995) *Flamenco*, a gene controlling the *gypsy* retrovirus of *Drosophila melanogaster*. Genetics 139:697–711

Pélisson A, Sarot E, Payen-Groschêne G, Bucheton A (2007) A novel repeat-associated small interfering RNA-mediated silencing pathway downregulates complementary sense *gypsy* transcripts in somatic cells of the *Drosophila* ovary. J Virol 81:1951–1960

Rohrmann GF, Karplus PA (2001) Relatedness of baculovirus and *gypsy* retrotransposon envelope proteins. BMC Evol Biol 1:1

Sarot E, Payen-Groschêne G, Bucheton A, Pélisson A (2004) Evidence for a piwi-dependent RNA silencing of the *gypsy* endogenous retrovirus by the *Drosophila melanogaster* *flamenco* gene. Genetics 166:1313–1321

Song SU, Kurkulos M, Boeke JD, Corces VG (1997) Infection of the germ line by retroviral particles produced in the follicle cells: a possible mechanism for the mobilization of the *gypsy* retroelement of *Drosophila*. Development 124:2789–2798

Song SU, Gerasimova T, Kurkulos M, Boeke JD, Corces VG (1994) An env-like protein encoded by a *Drosophila* retroelement: evidence that *gypsy* is an infectious retrovirus. Genes Dev 8:2046–2057

Tanda S, Mullor JL, Corces VG (1994) The *Drosophila* tom retrotransposon encodes an envelope protein. Mol Cell Biol 14:5392–5401

Temin HM (1980) Origin of retroviruses from cellular moveable genetic elements. Cell 21:599–600

Terzian C, Ferraz C, Demaille J, Bucheton A (2000) Evolution of the *Gypsy* endogenous retrovirus in the *Drosophila melanogaster* subgroup. Mol Biol Evol 17:908–914

Teysset L, Burns JC, Shike H, Sullivan BL, Bucheton A, Terzian C (1998) A Moloney murine leukemia virus-based retroviral vector pseudotyped by the insect retroviral *gypsy* envelope can infect *Drosophila* cells. J Virol 72:853–856

Vázquez-Manrique RP, Hernández M, Martínez-Sebastián MJ, De Frutos R (2000) Evolution of *gypsy* endogenous retrovirus in the *Drosophila* obscura species group. Mol Biol Evol 17:1185–1193

Weissenhorn W, Hinz A, Gaudin Y (2007) Virus membrane fusion. FEBS Lett 581:2150–2155

Westenberg M, Veenman F, Roode EC, Goldbach RW, Vlak JM, Zuidema D (2004) Functional analysis of the putative fusion domain of the baculovirus envelope fusion protein F. J Virol 78:6946–6954

Genome Dyn Stab (4)
D.-H. Lankenau, J.-N. Volff: Transposons and the Dynamic Genome
DOI 10.1007/7050_2008_042/Published online: 22 October 2008
© Springer-Verlag Berlin Heidelberg 2008

Transposon–Host Cell Interactions in the Regulation of *Sleeping Beauty* Transposition

Oliver Walisko[1,2] · Tobias Jursch[1] · Zsuzsanna Izsvák[1] · Zoltán Ivics[1] (✉)

[1]Max Delbrück Center for Molecular Medicine, Robert Rössle Str. 10, 13092 Berlin, Germany
zivics@mdc-berlin.de

[2]*Present address:*
Laboratoire de Microbiologie et Génétique Moléculaire, 31062 Toulouse, France

Abstract The mobility of transposable elements in natural populations is usually strictly regulated in order to preserve genomic stability. It is believed that the fine control of transposon movement is brought about by both transposon- and host cell-encoded factors and mechanisms. The *Sleeping Beauty* (SB) transposon is a reconstructed element, the first ever shown to be active in any vertebrate-derived cell. SB not only represents a powerful gene vector system for genomic manipulations in vertebrate species, but also has been serving as a useful experimental system to address transposon–host cell interactions at the molecular level. We have established that, in addition to the element-encoded transposase, cellular factors are involved in SB transposition and its regulation. Here we review regulatory mechanisms affecting transposition, with a special emphasis on only those that have been described for the SB element. Regulatory processes that act on the levels of transcription, chromatin, cell-cycle regulation, double-strand DNA break repair, and target site selection will be presented.

Abbreviations

DSB	Double-strand break
HDR	Homology-dependent repair
HIV	Human immunodeficiency virus
HMG	High-mobility group
IAP	Intracisternal A-particle
IHF	Integration host factor
IS	Insertion sequence
LINE	Long interspersed nuclear element
MLV	Murine leukemia virus
Mu	Mutator
NHEJ	Nonhomologous end-joining
SB	*Sleeping Beauty* transposon
TIR	Terminal inverted repeat
UTR	Untranslated region
V(D)J	Variable (diversity) joining

1
Introduction

The term "mobile DNA" designates genetic elements that are, unlike sequences solely devoted to producing proteins (protein-coding genes), capable of changing their location in the genome in a process generally called transposition. Because the ability to move appears to serve exclusively the benefit of these elements without providing any advantage for the host, mobile DNA is often referred to as "parasitic" or "selfish" DNA. However, transposable elements also represent an evolutionary force with the potential to provide genome plasticity under unfavorable environmental conditions that may lead to the emergence of altered gene expression and resulting adaptation to the new environmental challenges (McClintock 1984). Broadly, transposable elements fall into two classes, one of which transposes through an RNA intermediate (retrotransposons) and one directly transposes as DNA (transposons). Transposons and retrotransposons form a substantial fraction of the genomes of many organisms. For example, 44% of the human genome consists of transposable elements (Lander et al. 2001) of which 41% are derived from retrotransposons and 3% are transposon-derived. Given this large number, it is not surprising that transposable elements have shaped the structure and influenced the function of genomes, and that they have evolved together with the genome as an indwelling component. This coevolution gave rise to intimate transposon–host interactions, which we define as processes that involve components of both the transposable element and the host genome to limit transposable element-induced genome damage but also allow the element to move.

2
The *Sleeping Beauty* Transposable Element: Structure and Mechanism of Transposition

Sleeping Beauty (SB), an ancient, reconstructed transposon from fish (Ivics et al. 1997), is a member of the large Tc1/*mariner* superfamily of transposable elements (Plasterk et al. 1999). SB contains a single gene encoding the transposase flanked by two terminal inverted repeats (TIRs) (Fig. 1). SB transposition is thought to be a highly regulated process, initiated by the binding of the transposase to sites that are located in the TIRs (Fig. 1). The ends of the element are then probably paired through interaction of the transposase subunits, thereby forming a synaptic complex in the context of which the element gets excised from its donor locus (Fig. 1). The transposition reaction is completed after reintegration of the element into a new target site, followed by the repair of transposon-induced DNA lesions by the host repair machinery (Fig. 1). Each of these steps will be discussed in detail in subsequent sections of this review.

SB can efficiently transpose in cells of different vertebrate classes in tissue culture (Izsvák et al. 2000) as well as in somatic and germline tissues of fish, frogs, mice, and rats in vivo (Yant et al. 2000; Dupuy et al. 2001; Fischer et al. 2001; Davidson et al. 2003; Horie et al. 2003; Balciunas et al. 2004; Geurts et al. 2006a; Sinzelle et al. 2006; Kitada et al. 2007; Lu et al. 2007; Takeda et al. 2007; Yergeau and Mead 2007). Thus, SB is receiving considerable interest as a vector platform for the development of insertional mutagenesis screens in model organisms (reviewed in Miskey et al. 2005; Mátés et al. 2007) as well as for human therapeutic applications (reviewed in Izsvák and Ivics 2004; Essner et al. 2005; Hackett et al. 2005; Ivics and Izsvák 2006). SB also lends itself as a useful experimental system to address questions related to the basic molecular mechanisms of DNA transposition and its regulation in vertebrate cells.

3
Regulation of Transposition

De novo transposition events only become evolutionarily manifested, i.e., they only survive, if they can be stably transmitted to the next generation. Hence, restriction of transposition events to the germline is thought to ensure that new transposon insertions are inherited by the next generation, and to avoid evolutionarily unproductive (but still potentially mutagenic) events in somatic tissues.

A prototypic example for confining transposition events to the germline is provided by the P element transposon in *Drosophila melanogaster*. Expression of the active P element transposase protein is restricted to the germline by tissue-specific, selective splicing of the transposase messenger RNA (Laski et al. 1986). For the I factor, a *Drosophila* non-LTR LINE retrotransposon, expression of the ORF1 protein, which is essential for transposition, is limited to germline cells, where transposition occurs at high frequencies (Seleme et al. 1999). Expression of the intracisternal A-particle (IAP) LTR retrotransposon was also shown to be restricted to the male germline in mice (Dupressoir and Heidmann 1996). Similarly, expression of RNA and proteins associated with L1 (the major non-LTR retrotransposon in humans) preferentially occurs in germ cells (Branciforte and Martin 1994; Trelogan and Martin 1995; Ergun et al. 2004). Although the germline appears to be an attractive environment in which the products of transposition can be passed on to future generations, this strategy can be counteracted by protective mechanisms evolved by the host. For example, Tc1/*mariner* elements in the nematode *Caenorhabditis elegans* are active in the soma but silenced in the germline (Emmons and Yesner 1984; Eide and Anderson 1988) by RNA interference (Sijen and Plasterk 2003), which most likely protects the genome from heritable, transposition-generated defects.

Fig. 1 Mechanism and regulation of *Sleeping Beauty* transposition. **a** Overview of the ▶
transposition reaction. The transposable element consists of a gene encoding a transposase bracketed by terminal inverted repeats (*solid black arrows*) that contain binding sites of the transposase (*white arrows*) and flanking donor DNA. Transcriptional control elements in the 5′-UTR of the transposon drive transcription (*arrow*) of the transposase gene. The transposase (*spheres*) binds to its sites within the transposon inverted repeats. Excision takes place in a synaptic complex, and separates the transposon from the donor DNA. The excised element integrates into a TA site in the target DNA that will be duplicated and will be flanking the newly integrated transposon. On the right, the various steps of transposition are shown. On the left, mechanisms and host factors regulating each step of the transposition reaction are indicated. **b** Molecular events during cut-and-paste transposition. The transposase initiates the excision of the transposon with staggered cuts and reintegrates it at a TA target dinucleotide. The single-stranded gaps at the integration site as well as the double-strand DNA breaks in the donor DNA are repaired by the host DNA repair machinery. After repair, the target TA is duplicated at the integration site, and a small footprint is left behind at the site of excision. The footprint is an indication for an illegitimate NHEJ DSB repair mechanism

Irrespective of the tissue in which transposition occurs, one definitive consequence of a completed transposition event is that a copy of the transposable element has been inserted into a new location somewhere in the host genome. This inherent quality of mobile DNAs to insert themselves into the host DNA constitutes a potential threat to the overall integrity of the host genome. For example, insertions close to or within genes may lead to misexpression due to transcriptional up- or downregulation, and insertions into introns may result in altered splicing patterns, whereas insertions into exons may give rise to loss-of-function mutations. Furthermore, the accumulation of repeated sequences may lead to increased irregular recombination events and to genome instability. Therefore, it is of existential interest of the element to minimize damage to the integrity of the host genome. This is because an insertion event that in the worst case kills the host organism will consequently not be beneficial for the transposable element either, since its fate is intimately linked to the host. In evolutionary terms, this close relationship has resulted in multiple layers of regulation manifested at many levels of the various biochemical steps of the transposition process.

3.1
Transcriptional Control of Transposition

Expression of factors required for the transposition process is a limiting step in the transposition reaction, and therefore constitutes a major checkpoint in the transposition process (Fig. 1). Transposase expression from endogenous promoters requires host-encoded factors, such as RNA polymerases and accessory proteins, and hence represents an important interface between the transposable element and the host organism. In general, endogenous promoters appear to drive transposase expression rather inefficiently, which is

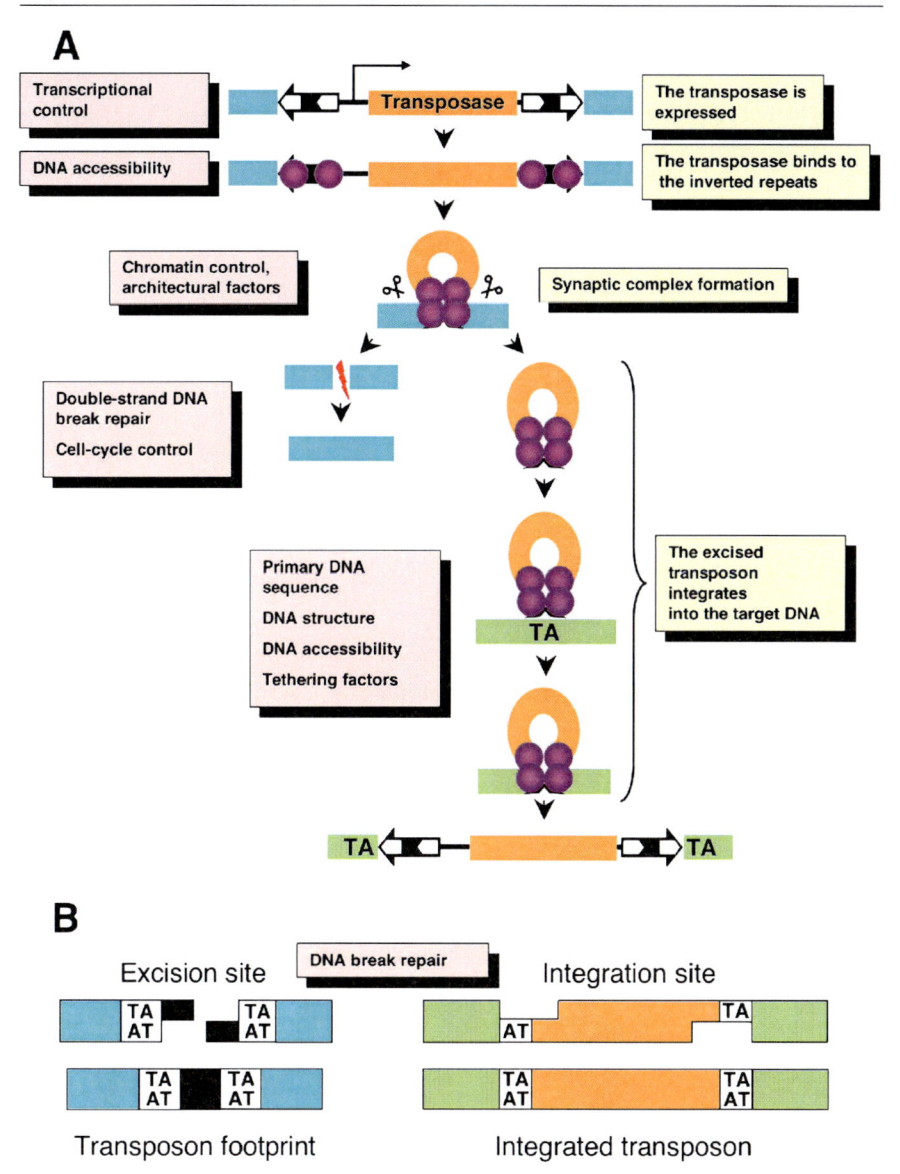

exemplified by the bacterial insertion sequences IS*911*, IS*21* and IS*30* and the maize *Spm* and *Ac* transposons (Dalrymple and Arber 1985; Kunze et al. 1987; Reimmann et al. 1989; Raina et al. 1993; Duval-Valentin et al. 2001). Endogenous promoters are located in the TIRs, which in addition contain the transposase binding sites. This permits regulation of promoter activity by the transposase or its truncated derivatives, because transposase binding at the TIRs may partially block access of transcription factors to the pro-

moter. This has been shown, for example, for the bacterial insertion sequence
IS911 (Duval-Valentin et al. 2001) and for the eukaryotic P element transposon (Kaufman and Rio 1991). For the P element, the transposase was shown
to prevent assembly of the RNA polymerase II complex in vitro (Kaufman and
Rio 1991).

DNA methylation is an effective means of transcriptional gene regulation, and it was suggested to serve as a host-defense mechanism to restrict
transposable element mobility (Yoder et al. 1997). Indeed, transcription of
the transposase gene can be negatively affected by DNA methylation of the
promoter region. For example, IS10 and Tn5 elements contain DNA methylation sites close to or overlapping the endogenous promoters; methylation
of these sites leads to inefficient transcription (Roberts et al. 1985; Yin et al.
1988). Similar findings were also obtained for eukaryotic transposable elements: transcription of the maize Ac transposon is abolished when Ac DNA is
methylated (Kunze et al. 1988), the promoter of the LINE-1 (L1) retrotransposon is repressed when methylated (Thayer et al. 1993; Hata and Sakaki
1997; Yu et al. 2001), and the mouse endogenous retrovirus IAP gets transcriptionally activated in cells deficient in DNA methyltransferase (Walsh
et al. 1998). Hypomethylation of DNA in Arabidopsis leads to activation of
both DNA transposons and retrotransposons (Hirochika et al. 2000; Miura
et al. 2001). In addition to transcriptional silencing, methylation of cytosine
residues leads to deamination and thus results in rapid sequence divergence.

The activity of promoter sequences located within transposable elements are frequently found to be regulated by host-encoded proteins, which
most likely allows spatial and temporal regulation of recombinase expression to be imposed. For example, for the L1 retrotransposon, a relatively
large set of host-encoded proteins have been identified to bind to the 5'-
untranslated region (UTR), which contains an internal promoter (Lavie et al.
2004). A particularly significant group of proteins are the HMG-box transcription factors.[1] Two members of the HMG-box family of transcription
factors have been found to interact with the 5'-UTR of L1: SOX11, which is
a positive regulator of L1 transcription upon overexpression (Tchenio et al.
2000), and SOX2, which represses L1 protein expression (Muotri et al. 2005).

---

[1] High-mobility group (HMG) proteins are classified into three subfamilies, HMGB1/2 (formerly
known as HMG1/2), HMGA1a/b (formerly known as HMGI/Y), and HMGN1/2 (formerly known
as HMGB14/17), which share many physical characteristics but differ in their main functional domains (Thomas and Travers 2001). Both the HMGB and HMGA1 group proteins are known to
bind A/T-rich DNA through interactions with the minor groove of the DNA helix (Bustin 1999).
HMGB1 is an abundant ($\sim 10^6$ molecules/cell), non-histone, nuclear protein associated with eukaryotic chromatin. Through its DNA-binding domain, termed the HMG box, HMGB1 binds DNA
in a sequence-independent manner, but with preference for certain DNA structures including four-
way junctions and severely undertwisted DNA (Bianchi et al. 1989; Bianchi et al. 1992). HMGB1 has
low affinity to B-form DNA, and is thought to be recruited by other DNA-binding proteins through
protein–protein interactions, and induce a local distortion of the DNA upon binding. Together with
the closely related HMGB2 protein, HMGB1 has been implicated in a number of eukaryotic cellular
processes including gene regulation, DNA replication, and recombination (Bustin 1999; Muller et al.
2001).

Furthermore, the Ying Yang-1 (YY-1) transcription factor was reported to bind to the 5′-UTR and contribute to transcriptional regulation of L1 (Minakami et al. 1992; Becker et al. 1993), whereas the runt-domain transcription factor RUNX3 was shown to decrease L1 transcription (Yang et al. 2003).

The UTRs of the SB transposon have been shown to direct transcription of the transposase gene (Walisko et al. 2008) (Fig. 1). In fact, the 5′-UTR drives expression of the SB transposase at a level sufficient to detect transposition in a colony-forming transposition assay in HeLa cells. Transcription from the 5′-UTR of SB is upregulated by the host-encoded factor HMG2L1 (Walisko et al. 2008), an HMG-box DNA binding domain-containing protein that shares structural similarity with lymphocyte enhancer binding factor 1 (LEF-1), sex-determining region Y (SRY), and SRY-related HMG-box protein 4 (SOX4) transcription factors (Bewley et al. 1998). The SB transposase physically interacts with HMG2L1, and antagonizes its effect on transcriptional regulation of the transposase gene (Walisko et al. 2008), suggesting that natural transposase expression is under a negative feedback regulation. This model postulates a sensitive balance in the regulation of transposase expression that is calibrated by transposase concentrations in the cell: low concentrations allow more transposase to be made, whereas high concentrations lead to shutting off transposase expression.

3.2
Control of Synaptic Complex Assembly During Transposition

The mobility of transposable elements is restricted by the requirement to form nucleoprotein complexes, which are a prerequisite for the execution of the chemical steps needed for a successful transposition event. Such structures, also called synaptic complexes or transpososomes, contain the DNA of the transposable element, the element-encoded recombinase(s), in some cases the target DNA and in many cases host-encoded protein factors that aid the formation and the stability of these complexes. Formation of such higher-order nucleoprotein complexes, which contain all DNA sites and protein components needed for the transposition reaction, protects cells from aberrant transposition events. Therefore, synaptic complex formation is an important checkpoint in transposition (Fig. 1), and there are examples emerging where host-encoded proteins contribute to this process, making transposition a joint work of two partners. For example, Mu, a temperate phage discovered in *Escherichia coli*, carries a linear, double-stranded genome, which can transpose into the bacterial genome. The structural organization of Mu is rather complex: each transposon end carries multiple MuA transposase binding sites, which show unequal orientation and spacing. At an early step of Mu transposition in vitro, when the left and the right transposon ends are brought together, the bacterial HU protein, a sequence-independent DNA-binding and bending protein, is required in addition to the MuA transposase

(Craigie and Mizuuchi 1987; Surette et al. 1987). HU binds to the left Mu end (Lavoie and Chaconas 1993), where it is thought to play an architectural role, which promotes synaptic complex assembly. In addition to the MuA transposase binding sites, the Mu transposon carries a transpositional enhancer sequence, which contains an integration host factor (IHF)-binding site. Like HU, IHF is a host-encoded, DNA-binding and bending protein (Robertson and Nash 1988), which stimulates transposition in vitro by acting on the enhancer sequence through the introduction of a sharp bend (Surette et al. 1989) that plays an important role in synaptic complex assembly. IHF was also found to modulate Tn10 transposition both in vitro (Chalmers et al. 1998) and in vivo (Signon and Kleckner 1995) by binding adjacent to the outside end of Tn10. The IHF-bent end of Tn10 then wraps around the transposase to form an activated synaptic complex (Crellin and Chalmers 2001).

The concept of transpositional regulation through formation of nucleoprotein complexes can also be found in vertebrate systems. For example, the variable (diversity) joining (V(D)J) recombination system that evolved from a DNA transposable element (Agrawal et al. 1998; Hiom et al. 1998; Kapitonov and Jurka 2005), and has been "domesticated" (Miller et al. 1992) to carry out recombination reactions to generate immunoglobulin and T-cell receptor gene diversity (Jones and Gellert 2004), requires host-encoded factors for synaptic complex assembly and successful excision. During V(D)J recombination, immunoglobulin and T-cell receptor genes are assembled from preexisting gene segments, which are separated by so-called recombination signal sequences (RSSs). Each RSS consists of conserved heptamer and nonamer regions, which are separated by a relatively nonconserved spacer region of 12 or 23 bp (12-RSS and 23-RSS). Binding and cleavage by the RAG1/2 recombinase is more efficient at the 12-RSS than at the 23-RSS, but both binding and cleavage of the 23-RSS can be enhanced by the high-mobility group proteins 1 and 2 (HMGB1/2) (van Gent et al. 1997). The recombination activating gene product RAG1 physically interacts with HMGB1/2 (Aidinis et al. 1999), which facilitates binding of RAG1 to the 23-RSS by bridging the distance in the 23-RSS between the heptamer and the nonamer regions through a sharp bend in the DNA (van Gent et al. 1997). HMGB1 has also been found to enhance integration of avian sarcoma virus (Aiyar et al. 1996). Another member of the HMG protein family, HMGI(Y), has been found to be associated with the viral preintegration complex of human immunodeficiency virus type 1 (HIV-1) and murine leukemia virus (MLV), a large nucleoprotein complex competent for integration (Farnet and Bushman 1997; Li et al. 2000). Binding of HMGI(Y) to the viral cDNA aids compaction of the viral DNA to form an integration-competent complex (Li et al. 2000).

From the Tc1/*mariner* transposon superfamily, the best characterized system in terms of synaptic complex assembly is the *Mos1 mariner* element originally discovered in *Drosophila mauritiana* (Hartl 2001). The Mos1 transposase binds differentially to the imperfect TIRs of the element (Zhang et al.

2001) with a higher affinity to the right end, where cleavage preferentially occurs. The Mos1 transposase first cleaves the nontransferred strand (the 5′-end of the transposon) within a single-end complex before the two ends are juxtaposed to form a paired-end complex (PEC), in which cleavage of the transferred strand (the 3′-end of the transposon) occurs (Dawson and Finnegan 2003). Transposase mutants that result in a reduced transposase–transposase interaction mobilize the element with reduced activity, suggesting that transposase–transposase interactions are required to form a catalytically active PEC (Zhang et al. 2001). Whereas crystallographic studies of the Mos1 transposase suggest that the PEC contains the two DNA ends together with a transposase dimer (Richardson et al. 2006; Richardson et al. 2007), biochemical studies proposed that the two ends are brought together by a transposase tetramer (Auge-Gouillou et al. 2005a; Auge-Gouillou et al. 2005b).

In *Sleeping Beauty* transposition, HMGB1 physically interacts with the SB transposase, which then recruits HMGB1 to its binding sites located in the TIRs, where HMGB1 can affect local DNA structure. HMGB1 enhances transposase binding to the TIRs, or stabilizes already formed complexes. Increased transpositional activity of the SB transposable element in the presence of HMGB1 is most likely due to a role of HMGB1 in synaptic complex formation, which involves the two TIRs held together by four transposase molecules (Izsvák et al. 2002; Zayed et al. 2003). HMGB1's effect on transposase binding is more pronounced at internal binding sites, suggesting that HMGB1's role in the transposition process is to ensure ordered assembly of a synaptic complex.

3.3
Regulation of Transposition by Chromatin

Formation of a catalytically active synaptic complex requires expression of the recombinase specific for the element and a DNA topology, which makes the element accessible for the protein machinery required for catalysis. The eukaryotic genome is typically organized into either of two types of chromatin: euchromatin, a relatively relaxed chromatin structure, in which the DNA is packed less tightly, and heterochromatin, a more inaccessible and highly condensed fraction of the genome. Heterochromatic regions carry characteristic features, which distinguish them from euchromatic DNA, such as dense cytosine methylation (5-Me-C), hypoacetylation of lysine residues in the N-terminal tails of histone H3 and H4, and methylation of specific lysine residues such as lysine 9 in histone H3. In contrast to euchromatin, which is largely composed of unique (protein coding) sequences, the DNA sequence of heterochromatin is usually repetitive and gene poor (Grewal and Jia 2007). One class of repetitive sequences found in heterochromatic regions of different genomes are transposable elements, and therefore it is believed that the accumulation of transposable DNA sequences in heterochromatic regions provides a "safe" place, where the deleterious potential of these elements can be kept on a leash (Bender 2004). Indeed, there is a strong correlation between chromatin structure and the activity of transposable elements. For example, insertion of reporter genes in or in close proximity to heterochromatin results in silencing of gene expression (Thon and Klar 1992; Allshire et al. 1994; Ayoub et al. 1999; Lahue et al. 2005), suggesting that heterochromatin represses transcription of genes located within or nearby. Thus, recruiting transposable DNAs into heterochromatic regions may provide efficient silencing of transcription of element-encoded proteins, and thus provides genome stability. In addition to its repressive function on transcription, heterochromatin also exerts a repressive influence on recombination (Clarke et al. 1986; Nakaseko et al. 1986) (compare also May, Slingsby, and Jeffreys, this SERIES; and Lichten, this SERIES); hence, containing repeated sequences in heterochromatic regions may prevent irregular recombination and genome instability.

One host mechanism regulating SB transposition is CpG methylation, and subsequent heterochromatin formation at the transposon donor site (Yusa et al. 2004). Unexpectedly, methylated SB transposons from both episomal and genomic contexts excise approximately 100-fold more efficiently than unmethylated transposons (Yusa et al. 2004). Chromatin immunoprecipitation experiments revealed that the hyperactive genomic donor sites have the characteristics of a heterochromatic structure. The SB transposase was found to colocalize with heterochromatin protein one (HP1), a well-established marker for heterochromatin, suggesting the transposase preferentially associates with heterochromatic DNA (Ikeda et al. 2007). Two models have been put forward to explain the effect of CpG methylation on SB transposition (Yusa et al. 2004). The first model involves interaction of yet unidentified cellular factors with methylated DNA and/or with DNA-bound transposase, resulting in enhanced transposition. The second model proposes that formation of heterochromatin at the transposon inverted repeats might facilitate the formation of a catalytically active synaptic complex (Fig. 2). Thus, similarly to the effect of HMGB1, conformational changes of the excising transposon may greatly influence the efficiency of transposition (Fig. 2).

3.4
Regulation by Cell-Cycle and DNA Repair Processes

The gap phases G1 and G2 provide important checkpoints in the cell cycle of proliferating cells, at which the presence of damaged DNA is detected, and sufficient time is allocated for repair prior to the onset of DNA replication and mitosis. DNA double-strand breaks (DSBs) represent the most hazardous type of DNA damage that, if left unrepaired, can lead to genomic instability. DSBs can be introduced not only by exogenous agents, such as ionizing

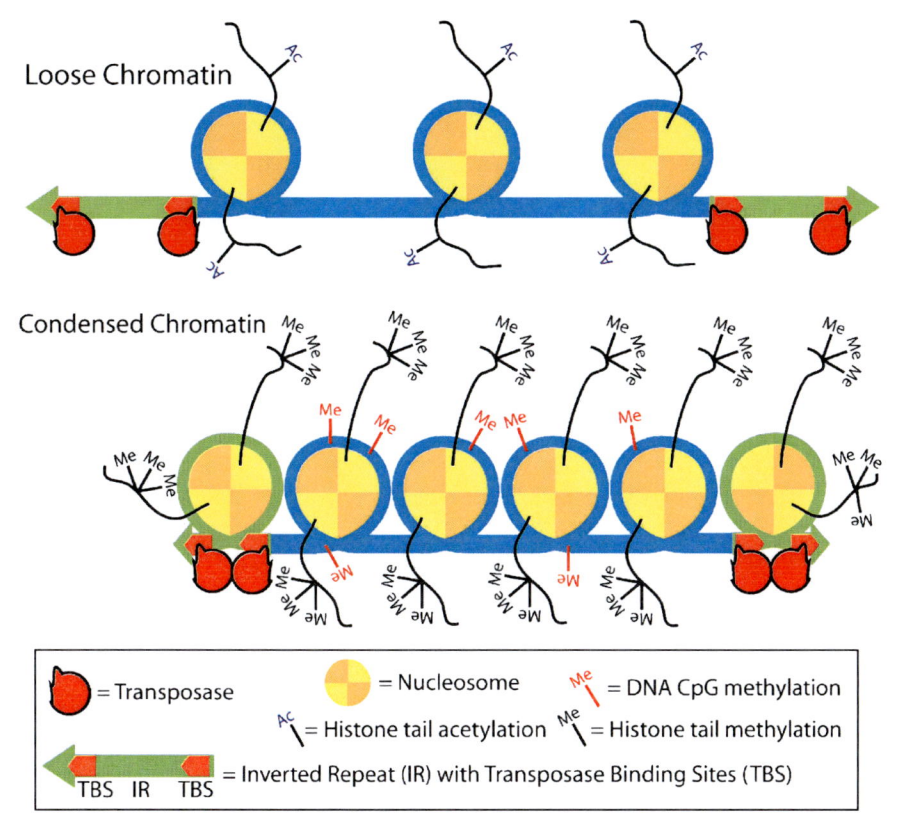

Fig. 2 A model for the enhancing effect of a compact chromatin structure on *Sleeping Beauty* transposition. Euchromatin contains DNA wrapped around nucleosomes in a "beads-along-a-string"-like conformation (*upper panel*). Transposase subunits bound within the transposon inverted repeats (IRs) are separated by 166-bp DNA. Heterochromatin (*lower panel*), characterized by DNA CpG methylation and specific histone tail modifications, e.g., trimethylated lysine 9 of histone H3, features a higher histone : DNA ratio. Positioning of a nucleosome between the transposase binding sites (TBSs) will shorten the distance between these sites, and could facilitate the formation of transposase dimers and subsequent assembly of the synaptic complex

radiation or chemicals, but also by endogenous cellular processes, such as DNA transposition or V(D)J recombination. Due to their possibly fatal consequences, pathways have evolved that allow the efficient repair of DSBs (Izsvák, Wang, and Ivics, this BOOK). In nonhomologous end-joining (NHEJ), the broken ends of DNA are rejoined without a requirement for a homologous template. In contrast, for homology-dependent repair (HDR), extensive homology is required between the region with the DSB and a template (usually a sister chromatid or a homologous chromosome). The two pathways act at different stages of the cell cycle, NHEJ acting primarily during G1/early S (Lee et al. 1997) and HDR being active in late S/G2 (Takata et al. 1998).

V(D)J recombination is strictly dependent on the NHEJ pathway for the repair of RAG-mediated DSBs (Jackson and Jeggo 1995), because only these repair products yield new, potentially contiguous reading frames of immunoglobulin and T-cell receptor genes. To confine V(D)J recombination to the G1 phase of the cell-cycle, the RAG2 protein, which constitutes the recombinase (transposase) together with RAG1, accumulates during G1, declines before the cell enters S phase, and remains low throughout the rest of the cell-cycle (Lin and Desiderio 1994). Phosphorylation of RAG2 by the cyclin A-cdk2 complex shortens the half-life of the protein at the G1/S boundary, whereas overexpression of p27^{Kip1}, a negative regulator of cyclin A-cdk2, results in elevated levels of RAG2, G1 arrest, and increased recombination (Lee and Desiderio 1999). Hence, cell-cycle regulated protein stability of RAG-2 confines V(D)J recombination to G1.

Many viruses have developed strategies to modulate the host cell-cycle machinery and cellular self-destruction mechanisms to maximize the chance for successful infection and the production of virus progeny. For example, the *vpr* accessory gene of HIV-1 blocks cellular proliferation at the G2 phase in various eukaryotic cells including T cells (He et al. 1995; Jowett et al. 1995), experimental cell lines such as HeLa or 293 cells (He et al. 1995; Re et al. 1995), or even yeast (Zhao et al. 1996), suggesting that the molecular mechanisms leading to *vpr*-induced G2 arrest are highly conserved. In addition to changes in the state of phosphorylation and subcellular compartmentalization of key cell-cycle regulatory proteins, *vpr*-induced herniations in the nuclear envelope and defects in the nuclear lamina have been proposed to contribute to the cell-cycle arrest (de Noronha et al. 2001). *Vpr* might also delay or prevent apoptosis of infected cells, thereby maximizing viral expression and increasing the amount of virus each infected cell produces (Goh et al. 1998). In addition, other viruses including herpes simplex virus (Lomonte and Everett 1999), cytomegalovirus (Lu and Shenk 1999), and Epstein–Barr virus (Cayrol and Flemington 1996) slow down the G1/S transition phase to allow ample opportunity for expression of viral genes before the onset of cellular genomic replication. Thus, overriding the normal cell-cycle program seems to be a shared strategy of many molecular parasites.

As stated above, transposons have evolved together with the genome as an indwelling component. However, similar to viruses, transposons are, for pragmatic reasons, best viewed as molecular parasites that propagate themselves using resources of the host cell. Recent evidence shows that transposable elements have also evolved mechanisms to modulate cell-cycle progression for their own benefit (Walisko and Ivics 2006; Walisko et al. 2006). Differential gene expression analysis with microarray hybridization revealed a significant decrease in cyclin D1 expression in human cells expressing the SB transposase. Downregulation of cyclin D1 results in a prolonged G1 phase of the cell-cycle and retarded growth of transposase-expressing cells. G1 slowdown is associated with a decrease of cyclin D1/cdk4-specific phosphorylation of the retinoblastoma protein. The Miz-1 transcription factor (Peukert et al. 1997) was identified as an interactor of the SB transposase in a yeast two-hybrid screen, and was found to be required for cyclin D1 downregulation by the transposase (Walisko et al. 2006). A temporary arrest of cells in G1 was found to enhance transposition, and a model has been proposed that links SB transposition to the G1 phase of the cell-cycle. The likely biological significance of this is that by inducing a temporary G1 delay, the SB transposase potentiates the involvement of NHEJ to repair transposition-inflicted DNA damage (Izsvák, Wang, and Ivics, this BOOK). Supporting this model, NHEJ factors are required for efficient SB transposition (Izsvák et al. 2004). Furthermore, the SB transposase physically interacts with the Ku DNA-binding subunit of DNA-PK, a key component of the NHEJ machinery (Izsvák et al. 2004). Likely aided by such interaction, DNA lesions generated by the excision of SB are predominantly repaired through the NHEJ pathway of DSB repair (Yant and Kay 2003; Izsvák et al. 2004). Thus, the transposase-induced G1 slowdown is probably a selfish act on the transposon's part to maximize the chance for a successful transposition event. It appears, therefore, that interference with the cell cycle is a shared strategy of parasitic genetic elements.

3.5
Target Site Selection and Integration

Some transposable elements lack a strong target site preference, and their distribution appears more or less random on the genomic scale (Halling and Kleckner 1982; Rosenzweig et al. 1983; Lee et al. 1987; Mori et al. 1988; Mizuuchi and Mizuuchi 1993; van Luenen and Plasterk 1994; Vigdal et al. 2002; Reznikoff 2003; Yant et al. 2005; Hackett et al. 2007). However, transposition is rarely fully random; for example, targeting of certain genomic sites may be specified by physical properties of the DNA, such as kinks due to protein binding, triplex DNA, or altered/abnormal DNA structures due to base composition, by the accessibility of DNA specified by local chromatin, as well as by preferential binding of proteins or protein complexes at certain sites. Yet other transposons are highly selective for the sites where they insert. A prototypic example for targeted transposition is the bacterial Tn7 transposon. Tn7 is highly specialized to insert into a single sequence motif in the *E. coli* genome called the attachment site of Tn7, *att*Tn7 (Lichtenstein and Brenner 1982), which is not deleterious for the host and a particular transposase subunit is responsible for targeting (Bainton et al. 1993). However, there are a number of examples in which transposable elements utilize host factors for targeted integration to genomic sites where insertions can be tolerated without apparent adverse effects. The yeast *Saccharomyces cerevisiae* Ty LTR retrotransposons exhibit a strong integration bias into "nonessential" genomic regions. For example, the Ty1 and Ty3 elements target upstream regions of RNA polymerase III (PolIII)-transcribed genes (Kirchner et al. 1995; Devine and Boeke 1996), which are proposed to provide a "safe haven" to tolerate insertions in densely packed genomes (Boeke and Devine 1998). Targeted integration of Ty3 requires the element-encoded integrase (IN) and PolIII transcription factors which tether the integration complex to the PolIII transcription machinery (Kirchner et al. 1995; Yieh et al. 2000; Yieh et al. 2002). Similar to Ty1/3, integration of the *Schizosaccharomyces pombe* Tf1 retrotransposon occurs into intergenic regions, but in contrast to Ty1/3, integration of Tf1 is biased for insertion into upstream regions of PolII-transcribed genes, which may be mediated by interactions between the IN and chromatin components (Behrens et al. 2000; Singleton and Levin 2002). Recently, an interaction between the Tf1 IN and the Atf1p transcription factor, as well as targeted Tf1 integration into promoter sequences mediated by Atf1p, was shown (Leem et al. 2008). The yeast Ty5 LTR retrotransposon shows a distinct integration specificity, which targets the element to transcriptionally silenced heterochromatic regions, such as telomeres and silent mating-type loci (Zou et al. 1996). This requires physical interaction of the IN and Sir4p, a structural component of silent chromatin (Xie et al. 2001). The eukaryotic microorganism *Dictyostelium discoideum* has a highly compact genome, and transposons in *D. discoideum* have developed two strategies to avoid genotoxic insertion into coding sequences. One of these strategies is nested integrations of transposons forming clusters. For example, the LTR retrotransposon family DIRS shows no initial target site selectivity, but can be found in a few clusters made up of several copies of itself (Loomis et al. 1995). The other strategy is targeted integration into "safe" regions of the genome away from protein-coding sequences. This strategy is primarily used by non-LTR retrotransposons that insert up- and downstream of tRNA genes (Winckler et al. 2005). The ORF1 protein encoded by the TRE5 element was recently shown to interact with TFIIIB, suggesting a role of this interaction in targeting integration into tRNA genes (Chung et al. 2007).

Unlike LTR retrotransposons that, due to the lack of the envelope gene, are restricted in their life cycle to an intracellular environment and hence are intrinsic components of the genome, retroviruses are proficient at existing in an extracellular state and replicate their genomes by infection. In addition, cells suffering from retroviral infections may persist only for a limited period of time before they are eliminated by the cytopathic effect of the infection or by immune clearance. This difference in their life cycle is manifested in a different targeting strategy of retroviruses as compared to retrotransposons. Analysis of integration site selection of HIV-1 revealed that integration occurs into transcriptional units of actively transcribed genes (Schroder et al. 2002). In contrast to HIV, MLV integration shows a different pattern; integration is less biased into transcription units as compared to HIV, but instead its preferred integration sites are near the start of transcriptional units (i.e., promoters) (Wu et al. 2003; Mitchell et al. 2004). A plausible explanation for the differences in target site selection observed between retrotransposons and retroviruses is that somatic cells infected by retroviruses may have a limited life span, and hence integration into transcriptionally active regions helps to ensure high transcription levels of viral DNA to maximize viral replication prior to the elimination of infected cells.

As discussed above, MLV and HIV show an integration bias toward transcriptionally active regions in the genome. Because no sequence-specific integration preference of the retroviral/lentiviral IN protein itself has been observed, biased genomic integration can be due to interaction of the viral components with certain host proteins or recognition of different chromatin states of the chromosomes during integration (Cherepanov et al. 2003; Mitchell et al. 2004). Indeed, the bias of HIV and HIV-derived lentiviral vectors toward integration into active cellular transcription units was proposed to be due to tethering interactions with cellular proteins rather than to chromatin accessibility. In particular, the cellular lens epithelium-derived growth factor (LEDGF/p75) was shown to influence HIV target site selection (Ciuffi and Bushman 2006). LEDGF/p75 acts as a transcriptional coactivator, and interacts with components of the basal transcription machinery (Ge et al. 1998) as well as with HIV IN via an integrase-binding domain (Vanegas et al. 2005).

Similar to most other transposable elements, SB does not integrate randomly into target DNA, and displays a certain degree of specificity in target site utilization at the primary DNA sequence level (Vigdal et al. 2002). Namely, SB exclusively integrates into TA dinucleotides that are duplicated upon transposition, and flank the integrated element (referred to as target site duplication) (Ivics et al. 1997; Vigdal et al. 2002) (Fig. 1). A palindromic AT-repeat consensus sequence associated with bendability and hydrogen bonding potential of the target DNA was found to affect SB's target selection (Vigdal et al. 2002). However, the primary sequence is clearly not the determining factor for target selection, because preferred integration sites ("hot spots") within a limited target DNA were found not to match the consensus sequence. Instead, it was shown that a unique deformation inherent in the sequence may be a recognition signal for target selection. This deformation, and therefore the likelihood a particular TA will be targeted by SB, can be predicted by the V(step) algorithm (Liu et al. 2005). Such predictions may allow us to assess theoretical risks associated with transposon insertions in particular genomic regions, an important parameter in the safety profile of SB-based gene vectors (Geurts et al. 2006b; Hackett et al. 2007).

Even though SB transposition shows considerable specificity at the actual DNA sequence level, analysis of a large number of both selected and unselected insertion sites recovered from both primary and cultured mammalian cells showed that SB integration can be considered fairly random on a genomic level (Vigdal et al. 2002; Yant et al. 2005). Roughly one-third of SB insertions occur in transcribed regions, suggesting no preference for or against insertion into genes. A tethering mechanism similar to that found for HIV IN and LEDGF/p75 has not been established for SB yet; however, since transposon insertion always takes place in the context of chromatin, DNA-binding proteins that interact with the transpositional machinery will likely have the capacity to tether the transpositional complex to certain loci. Supporting this expectation, an engineered fusion protein containing the tetracyclin repressor DNA-binding protein and N-57, a previously identified protein–protein interaction domain of the SB transposase (Izsvák et al. 2002), was shown to bias SB integration near tetracyclin operator sites at a frequency of > 10% in transfected human cells (Ivics et al. 2007).

4
Concluding Remarks

The first demonstrations of DNA transposition in cell-free reactions by using recombinant transposase proteins fueled ambitious expectations in some, that transposons will be able to jump in essentially any cell. By now, overwhelming evidence shows that, even though the transposase is sufficient to execute the biochemical steps of the transposition reaction, this process is regulated and modulated in a cellular environment. The transposase has to be expressed, it needs to interact with the transposon DNA, a complex in which excision takes place has to form, and the excised transposon needs to interact with the target DNA for insertion. In the context of a living cell, all of these processes are under regulation of the transcriptional apparatus, chromatin structure, and accessory proteins that influence DNA topology. Furthermore, factors that interact with the transposase and modulate cellular mechanisms, including the cell-cycle and DNA repair as well as DNA-binding proteins that tether the transpositional complex to certain chromosomal sites, all contribute to the outcome of the transposition reaction. SB has been, and will continue to be, a major experimental tool to elucidate many of these mechanisms.

Acknowledgements Work in the authors' laboratory was supported by a grant from the Deutsche Forschungsgemeinschaft (DFG) to Z.I. (IV 21/3-1).

References

Agrawal A, Eastman QM, Schatz DG (1998) Transposition mediated by RAG1 and RAG2 and its implications for the evolution of the immune system. Nature 394:744–751

Aidinis V, Bonaldi T, Beltrame M, Santagata S, Bianchi ME, Spanopoulou E (1999) The RAG1 homeodomain recruits HMG1 and HMG2 to facilitate recombination signal sequence binding and to enhance the intrinsic DNA-bending activity of RAG1–RAG2. Mol Cell Biol 19:6532–6542

Aiyar A, Hindmarsh P, Skalka AM, Leis J (1996) Concerted integration of linear retroviral DNA by the avian sarcoma virus integrase in vitro: dependence on both long terminal repeat termini. J Virol 70:3571–3580

Allshire RC, Javerzat JP, Redhead NJ, Cranston G (1994) Position effect variegation at fission yeast centromeres. Cell 76:157–169

Auge-Gouillou C, Brillet B, Germon S, Hamelin MH, Bigot Y (2005a) Mariner Mos1 transposase dimerizes prior to ITR binding. J Mol Biol 351:117–130

Auge-Gouillou C, Brillet B, Hamelin MH, Bigot Y (2005b) Assembly of the mariner Mos1 synaptic complex. Mol Cell Biol 25:2861–2870

Ayoub N, Goldshmidt I, Cohen A (1999) Position effect variegation at the mating-type locus of fission yeast: a cis-acting element inhibits covariegated expression of genes in the silent and expressed domains. Genetics 152:495–508

Bainton RJ, Kubo KM, Feng JN, Craig NL (1993) Tn7 transposition: target DNA recognition is mediated by multiple Tn7-encoded proteins in a purified in vitro system. Cell 72:931–943

Balciunas D, Davidson AE, Sivasubbu S, Hermanson SB, Welle Z, Ekker SC (2004) Enhancer trapping in zebrafish using the Sleeping Beauty transposon. BMC Genomics 5:62

Becker KG, Swergold GD, Ozato K, Thayer RE (1993) Binding of the ubiquitous nuclear transcription factor YY1 to a cis regulatory sequence in the human LINE-1 transposable element. Hum Mol Genet 2:1697–1702

Behrens R, Hayles J, Nurse P (2000) Fission yeast retrotransposon Tf1 integration is targeted to 5′ ends of open reading frames. Nucleic Acids Res 28:4709–4716

Bender J (2004) Chromatin-based silencing mechanisms. Curr Opin Plant Biol 7:521–526

Bewley CA, Gronenborn AM, Clore GM (1998) Minor groove-binding architectural proteins: structure, function, and DNA recognition. Annu Rev Biophys Biomol Struct 27:105–131

Bianchi ME, Beltrame M, Paonessa G (1989) Specific recognition of cruciform DNA by nuclear protein HMG1. Science 243:1056–1059

Bianchi ME, Falciola L, Ferrari S, Lilley DM (1992) The DNA binding site of HMG1 protein is composed of two similar segments (HMG boxes), both of which have counterparts in other eukaryotic regulatory proteins. EMBO J 11:1055–1063

Boeke JD, Devine SE (1998) Yeast retrotransposons: finding a nice quiet neighborhood. Cell 93:1087–1089

Branciforte D, Martin SL (1994) Developmental and cell type specificity of LINE-1 expression in mouse testis: implications for transposition. Mol Cell Biol 14:2584–2592

Bustin M (1999) Regulation of DNA-dependent activities by the functional motifs of the high-mobility-group chromosomal proteins. Mol Cell Biol 19:5237–5246

Cayrol C, Flemington EK (1996) The Epstein–Barr virus bZIP transcription factor Zta causes G0/G1 cell cycle arrest through induction of cyclin-dependent kinase inhibitors. EMBO J 15:2748–2759

Chalmers R, Guhathakurta A, Benjamin H, Kleckner N (1998) IHF modulation of Tn10 transposition: sensory transduction of supercoiling status via a proposed protein/DNA molecular spring. Cell 93:897–908

Cherepanov P, Maertens G, Proost P, Devreese B, Van Beeumen J, Engelborghs Y, De Clercq E et al (2003) HIV-1 integrase forms stable tetramers and associates with LEDGF/p75 protein in human cells. J Biol Chem 278:372–381

Chung T, Siol O, Dingermann T, Winckler T (2007) Protein interactions involved in tRNA gene-specific integration of Dictyostelium discoideum non-long terminal repeat retrotransposon TRE5-A. Mol Cell Biol 27:8492–8501

Ciuffi A, Bushman FD (2006) Retroviral DNA integration: HIV and the role of LEDGF/p75. Trends Genet 22:388–395

Clarke L, Amstutz H, Fishel B, Carbon J (1986) Analysis of centromeric DNA in the fission yeast Schizosaccharomyces pombe. Proc Natl Acad Sci USA 83:8253–8257

Craigie R, Mizuuchi K (1987) Transposition of Mu DNA: joining of Mu to target DNA can be uncoupled from cleavage at the ends of Mu. Cell 51:493–501

Crellin P, Chalmers R (2001) Protein–DNA contacts and conformational changes in the Tn10 transpososome during assembly and activation for cleavage. EMBO J 20:3882–3891

Dalrymple B, Arber W (1985) Promotion of RNA transcription on the insertion element IS30 of E. coli K12. EMBO J 4:2687–2693

Davidson AE, Balciunas D, Mohn D, Shaffer J, Hermanson S, Sivasubbu S, Cliff MP et al (2003) Efficient gene delivery and gene expression in zebrafish using the Sleeping Beauty transposon. Dev Biol 263:191–202

Dawson A, Finnegan DJ (2003) Excision of the Drosophila mariner transposon Mos1. Comparison with bacterial transposition and V(D)J recombination. Mol Cell 11:225–235

De Noronha CM, Sherman MP, Lin HW, Cavrois MV, Moir RD, Goldman RD, Greene WC (2001) Dynamic disruptions in nuclear envelope architecture and integrity induced by HIV-1 Vpr. Science 294:1105–1108

Devine SE, Boeke JD (1996) Integration of the yeast retrotransposon Ty1 is targeted to regions upstream of genes transcribed by RNA polymerase III. Genes Dev 10:620–633

Dupressoir A, Heidmann T (1996) Germ line-specific expression of intracisternal A-particle retrotransposons in transgenic mice. Mol Cell Biol 16:4495–503

Dupuy AJ, Fritz S, Largaespada DA (2001) Transposition and gene disruption in the male germline of the mouse. Genesis 30:82–88

Duval-Valentin G, Normand C, Khemici V, Marty B, Chandler M (2001) Transient promoter formation: a new feedback mechanism for regulation of IS911 transposition. EMBO J 20:5802–5811

Eide D, Anderson P (1988) Insertion and excision of Caenorhabditis elegans transposable element Tc1. Mol Cell Biol 8:737–746

Emmons SW, Yesner L (1984) High-frequency excision of transposable element Tc 1 in the nematode Caenorhabditis elegans is limited to somatic cells. Cell 36:599–605

Ergun S, Buschmann C, Heukeshoven J, Dammann K, Schnieders F, Lauke H, Chalajour F et al (2004) Cell type-specific expression of LINE-1 open reading frames 1 and 2 in fetal and adult human tissues. J Biol Chem 279:27753–27763

Essner JJ, McIvor RS, Hackett PB (2005) Awakening gene therapy with Sleeping Beauty transposons. Curr Opin Pharmacol 5:513–519

Farnet CM, Bushman FD (1997) HIV-1 cDNA integration: requirement of HMG I(Y) protein for function of preintegration complexes in vitro. Cell 88:483–492

Fischer SE, Wienholds E, Plasterk RH (2001) Regulated transposition of a fish transposon in the mouse germ line. Proc Natl Acad Sci USA 98:6759–6764

Ge H, Si Y, Roeder RG (1998) Isolation of cDNAs encoding novel transcription coactivators p52 and p75 reveals an alternate regulatory mechanism of transcriptional activation. EMBO J 17:6723–6729

Geurts AM, Collier LS, Geurts JL, Oseth LL, Bell ML, Mu D, Lucito R et al (2006a) Gene mutations and genomic rearrangements in the mouse as a result of transposon mobilization from chromosomal concatemers. PLoS Genet 2:e156

Geurts AM, Hackett CS, Bell JB, Bergemann TL, Collier LS, Carlson CM, Largaespada DA et al (2006b) Structure-based prediction of insertion-site preferences of transposons into chromosomes. Nucleic Acids Res 34:2803–2811

Goh WC, Rogel ME, Kinsey CM, Michael SF, Fultz PN, Nowak MA, Hahn BH et al (1998) HIV-1 Vpr increases viral expression by manipulation of the cell cycle: a mechanism for selection of Vpr in vivo. Nat Med 4:65–71

Grewal SI, Jia S (2007) Heterochromatin revisited. Nat Rev Genet 8:35–46

Hackett CS, Geurts AM, Hackett PB (2007) Predicting preferential DNA vector insertion sites: implications for functional genomics and gene therapy. Genome Biol 8(Suppl 1):S12

Hackett PB, Ekker SC, Largaespada DA, McIvor RS (2005) Sleeping Beauty transposon-mediated gene therapy for prolonged expression. Adv Genet 54:189–232

Halling SM, Kleckner N (1982) A symmetrical six-base-pair target site sequence determines Tn10 insertion specificity. Cell 28:155–163

Hartl D (2001) Discovery of the transposable element mariner. Genetics 157:471–476

Hata K, Sakaki Y (1997) Identification of critical CpG sites for repression of L1 transcription by DNA methylation. Gene 189:227–234

He J, Choe S, Walker R, Di Marzio P, Morgan DO, Landau NR (1995) Human immunodeficiency virus type 1 viral protein R (vpr) arrests cells in the G2 phase of the cell cycle by inhibiting p34cdc2 activity. J Virol 69:6705–6711

Hiom K, Melek M, Gellert M (1998) DNA transposition by the RAG1 and RAG2 proteins: a possible source of oncogenic translocations. Cell 94:463–470

Hirochika H, Okamoto H, Kakutani T (2000) Silencing of retrotransposons in Arabidopsis and reactivation by the ddm1 mutation. Plant Cell 12:357–369

Horie K, Yusa K, Yae K, Odajima J, Fischer SE, Keng VW, Hayakawa T et al (2003) Characterization of Sleeping Beauty transposition and its application to genetic screening in mice. Mol Cell Biol 23:9189–207

Ikeda R, Kokubu C, Yusa K, Keng VW, Horie K, Takeda J (2007) Sleeping Beauty transposase has an affinity for heterochromatin conformation. Mol Cell Biol 27:1665–1676

Ivics Z, Izsvák Z (2006) Transposons for gene therapy! Curr Gene Ther 6:593–607

Ivics Z, Hackett PB, Plasterk RH, Izsvák Z (1997) Molecular reconstruction of Sleeping Beauty, a Tc1-like transposon from fish, and its transposition in human cells. Cell 91:501–510

Ivics Z, Katzer A, Stuwe EE, Fiedler D, Knespel S, Izsvák Z (2007) Targeted Sleeping Beauty transposition in human cells. Mol Ther 15:1137–1144

Izsvák Z, Ivics Z (2004) Sleeping Beauty transposition: biology and applications for molecular therapy. Mol Ther 9:147–156

Izsvák Z, Ivics Z, Plasterk RH (2000) Sleeping Beauty, a wide host-range transposon vector for genetic transformation in vertebrates. J Mol Biol 302:93–102

Izsvák Z, Khare D, Behlke J, Heinemann U, Plasterk RH, Ivics Z (2002) Involvement of a bifunctional, paired-like DNA-binding domain and a transpositional enhancer in Sleeping Beauty transposition. J Biol Chem 277:34581–34588

Izsvák Z, Stuwe EE, Fiedler D, Katzer A, Jeggo PA, Ivics Z (2004) Healing the wounds inflicted by Sleeping Beauty transposition by double-strand break repair in mammalian somatic cells. Mol Cell 13:279–290

Jackson SP, Jeggo PA (1995) DNA double-strand break repair and V(D)J recombination: involvement of DNA-PK. Trends Biochem Sci 20:412–415

Jones JM, Gellert M (2004) The taming of a transposon: V(D)J recombination and the immune system. Immunol Rev 200:233–248

Jowett JB, Planelles V, Poon B, Shah NP, Chen ML, Chen IS (1995) The human immunodeficiency virus type 1 vpr gene arrests infected T cells in the G2 + M phase of the cell cycle. J Virol 69:6304–6313

Kapitonov VV, Jurka J (2005) RAG1 core and V(D)J recombination signal sequences were derived from Transib transposons. PLoS Biol 3:e181

Kaufman PD, Rio DC (1991) Drosophila P-element transposase is a transcriptional repressor in vitro. Proc Natl Acad Sci USA 88:2613–2617

Kirchner J, Connolly CM, Sandmeyer SB (1995) Requirement of RNA polymerase III transcription factors for in vitro position-specific integration of a retroviruslike element. Science 267:1488–1491

Kitada K, Ishishita S, Tosaka K, Takahashi R, Ueda M, Keng VW, Horie K et al (2007) Transposon-tagged mutagenesis in the rat. Nat Methods 4:131–133

Kunze R, Stochaj U, Laufs J, Starlinger P (1987) Transcription of transposable element activator (Ac) of Zea mays L. EMBO J 6:1555–1563

Kunze R, Starlinger P, Schwartz D (1988) DNA methylation of the maize transposable element Ac interferes with its transcription. Mol Gen Genet 214:325–327

Lahue E, Heckathorn J, Meyer Z, Smith J, Wolfe C (2005) The Saccharomyces cerevisiae Sub2 protein suppresses heterochromatic silencing at telomeres and subtelomeric genes. Yeast 22:537–551

Lander ES, Linton LM, Birren B, Nusbaum C, Zody MC, Baldwin J, Devon K et al (2001) Initial sequencing and analysis of the human genome. Nature 409:860–921

Laski FA, Rio DC, Rubin GM (1986) Tissue specificity of Drosophila P element transposition is regulated at the level of mRNA splicing. Cell 44:7–19

Lavie L, Maldener E, Brouha B, Meese EU, Mayer J (2004) The human L1 promoter: variable transcription initiation sites and a major impact of upstream flanking sequence on promoter activity. Genome Res 14:2253–2260

Lavoie BD, Chaconas G (1993) Site-specific HU binding in the Mu transpososome: conversion of a sequence-independent DNA-binding protein into a chemical nuclease. Genes Dev 7:2510–2519

Lee J, Desiderio S (1999) Cyclin A/CDK2 regulates V(D)J recombination by coordinating RAG-2 accumulation and DNA repair. Immunity 11:771–781

Lee SE, Mitchell RA, Cheng A, Hendrickson EA (1997) Evidence for DNA-PK-dependent and -independent DNA double-strand break repair pathways in mammalian cells as a function of the cell cycle. Mol Cell Biol 17:1425–1433

Lee SY, Butler D, Kleckner N (1987) Efficient Tn10 transposition into a DNA insertion hot spot in vivo requires the 5-methyl groups of symmetrically disposed thymines within the hot-spot consensus sequence. Proc Natl Acad Sci USA 84:7876–7880

Leem YE, Ripmaster TL, Kelly FD, Ebina H, Heincelman ME, Zhang K, Grewal SI et al (2008) Retrotransposon Tf1 is targeted to Pol II promoters by transcription activators. Mol Cell 30:98–107

Li L, Yoder K, Hansen MS, Olvera J, Miller MD, Bushman FD (2000) Retroviral cDNA integration: stimulation by HMG I family proteins. J Virol 74:10965–10974

Lichtenstein C, Brenner S (1982) Unique insertion site of Tn7 in the E. coli chromosome. Nature 297:601–603

Lin WC, Desiderio S (1994) Cell cycle regulation of V(D)J recombination-activating protein RAG-2. Proc Natl Acad Sci USA 91:2733–2737

Liu G, Geurts AM, Yae K, Srinivasan AR, Fahrenkrug SC, Largaespada DA, Takeda J et al (2005) Target-site preferences of Sleeping Beauty transposons. J Mol Biol 346:161–173

Lomonte P, Everett RD (1999) Herpes Simplex Virus type 1 immediate-early protein Vmw110 inhibits progression of cells through mitosis and from G1 into S phase of the cell cycle. J Virol 73:9456–9467

Loomis WF, Welker D, Hughes J, Maghakian D, Kuspa A (1995) Integrated maps of the chromosomes in Dictyostelium discoideum. Genetics 141:147–157

Lu B, Geurts AM, Poirier C, Petit DC, Harrison W, Overbeek PA, Bishop CE (2007) Generation of rat mutants using a coat color-tagged Sleeping Beauty transposon system. Mamm Genome 18:338–436

Lu M, Shenk T (1999) Human cytomegalovirus UL69 protein induces cells to accumulate in G1 phase of the cell cycle. J Virol 73:676–683

Mátés L, Izsvák Z, Ivics Z (2007) Technology transfer from worms and flies to vertebrates: transposition-based genome manipulations and their future perspectives. Genome Biol 8(Suppl 1):S1

McClintock B (1984) The significance of responses of the genome to challenge. Science 226:792–801

Miller WJ, Hagemann S, Reiter E, Pinsker W (1992) P-element homologous sequences are tandemly repeated in the genome of Drosophila guanche. Proc Natl Acad Sci USA 89:4018–4022

Minakami R, Kurose K, Etoh K, Furuhata Y, Hattori M, Sakaki Y (1992) Identification of an internal cis-element essential for the human L1 transcription and a nuclear factor(s) binding to the element. Nucleic Acids Res 20:3139–3145

Miskey C, Izsvak Z, Kawakami K, Ivics Z (2005) DNA transposons in vertebrate functional genomics. Cell Mol Life Sci 62:629–641

Mitchell RS, Beitzel BF, Schroder AR, Shinn P, Chen H, Berry CC, Ecker JR et al (2004) Retroviral DNA integration: ASLV, HIV, and MLV show distinct target site preferences. PLoS Biol 2:E234

Miura A, Yonebayashi S, Watanabe K, Toyama T, Shimada H, Kakutani T (2001) Mobilization of transposons by a mutation abolishing full DNA methylation in Arabidopsis. Nature 411:212–214

Mizuuchi M, Mizuuchi K (1993) Target site selection in transposition of phage Mu. Cold Spring Harb Symp Quant Biol 58:515–523

Mori I, Benian GM, Moerman DG, Waterston RH (1988) Transposable element Tc1 of Caenorhabditis elegans recognizes specific target sequences for integration. Proc Natl Acad Sci USA 85:861–864

Muller S, Scaffidi P, Degryse B, Bonaldi T, Ronfani L, Agresti A, Beltrame M et al (2001) The double life of HMGB1 chromatin protein: architectural factor and extracellular signal. EMBO J 20:4337–4340

Muotri AR, Chu VT, Marchetto MC, Deng W, Moran JV, Gage FH (2005) Somatic mosaicism in neuronal precursor cells mediated by L1 retrotransposition. Nature 435:903–910

Nakaseko Y, Adachi Y, Funahashi S, Niwa O, Yanagida M (1986) Chromosome walking shows a highly homologous repetitive sequence present in all the centromere regions of fission yeast. EMBO J 5:1011–1021

Peukert K, Staller P, Schneider A, Carmichael G, Hanel F, Eilers M (1997) An alternative pathway for gene regulation by Myc. EMBO J 16:5672–5686

Plasterk RH, Izsvák Z, Ivics Z (1999) Resident aliens: the Tc1/mariner superfamily of transposable elements. Trends Genet 15:326–332

Raina R, Cook D, Fedoroff N (1993) Maize Spm transposable element has an enhancer-insensitive promoter. Proc Natl Acad Sci USA 90:6355–6359

Re F, Braaten D, Franke EK, Luban J (1995) Human immunodeficiency virus type 1 Vpr arrests the cell cycle in G2 by inhibiting the activation of p34cdc2-cyclin B. J Virol 69:6859–6864

Reimmann C, Moore R, Little S, Savioz A, Willetts NS, Haas D (1989) Genetic structure, function and regulation of the transposable element IS21. Mol Gen Genet 215:416–424

Reznikoff WS (2003) Tn5 as a model for understanding DNA transposition. Mol Microbiol 47:1199–1206

Richardson JM, Dawson A, O'Hagan N, Taylor P, Finnegan DJ, Walkinshaw MD (2006) Mechanism of Mos1 transposition: insights from structural analysis. EMBO J 25:1324–1334

Richardson JM, Finnegan DJ, Walkinshaw MD (2007) Crystallization of a Mos1 transposase-inverted-repeat DNA complex: biochemical and preliminary crystallographic analyses. Acta Crystallogr Sect F Struct Biol Cryst Commun 63:434–437

Roberts D, Hoopes BC, McClure WR, Kleckner N (1985) IS10 transposition is regulated by DNA adenine methylation. Cell 43:117–130

Robertson CA, Nash HA (1988) Bending of the bacteriophage lambda attachment site by Escherichia coli integration host factor. J Biol Chem 263:3554–3557

Rosenzweig B, Liao LW, Hirsh D (1983) Target sequences for the C. elegans transposable element Tc1. Nucleic Acids Res 11:7137–7140

Schroder AR, Shinn P, Chen H, Berry C, Ecker JR, Bushman F (2002) HIV-1 integration in the human genome favors active genes and local hotspots. Cell 110:521–529

Seleme MC, Busseau I, Malinsky S, Bucheton A, Teninges D (1999) High-frequency retrotransposition of a marked I factor in Drosophila melanogaster correlates with a dynamic expression pattern of the ORF1 protein in the cytoplasm of oocytes. Genetics 151:761–771

Signon L, Kleckner N (1995) Negative and positive regulation of Tn10/IS10-promoted recombination by IHF: two distinguishable processes inhibit transposition off of multicopy plasmid replicons and activate chromosomal events that favor evolution of new transposons. Genes Dev 9:1123–1136

Sijen T, Plasterk RH (2003) Transposon silencing in the Caenorhabditis elegans germ line by natural RNAi. Nature 426:310–314

Singleton TL, Levin HL (2002) A long terminal repeat retrotransposon of fission yeast has strong preferences for specific sites of insertion. Eukaryot Cell 1:44–55

Sinzelle L, Vallin J, Coen L, Chesneau A, Du Pasquier D, Pollet N, Demeneix B et al (2006) Generation of trangenic Xenopus laevis using the Sleeping Beauty transposon system. Transgenic Res 15:751–760

Surette MG, Buch SJ, Chaconas G (1987) Transpososomes: stable protein–DNA complexes involved in the in vitro transposition of bacteriophage Mu DNA. Cell 49:253–262

Surette MG, Lavoie BD, Chaconas G (1989) Action at a distance in Mu DNA transposition: an enhancer-like element is the site of action of supercoiling relief activity by integration host factor (IHF). EMBO J 8:3483–3489

Takata M, Sasaki MS, Sonoda E, Morrison C, Hashimoto M, Utsumi H, Yamaguchi-Iwai Y et al (1998) Homologous recombination and non-homologous end-joining pathways of DNA double-strand break repair have overlapping roles in the maintenance of chromosomal integrity in vertebrate cells. EMBO J 17:5497–5508

Takeda J, Keng VW, Horie K (2007) Germline mutagenesis mediated by Sleeping Beauty transposon system in mice. Genome Biol 8(Suppl 1):S14

Tchenio T, Casella JF, Heidmann T (2000) Members of the SRY family regulate the human LINE retrotransposons. Nucleic Acids Res 28:411–415

Thayer RE, Singer MF, Fanning TG (1993) Undermethylation of specific LINE-1 sequences in human cells producing a LINE-1-encoded protein. Gene 133:273–277

Thomas JO, Travers AA (2001) HMG1 and 2, and related "architectural" DNA-binding proteins. Trends Biochem Sci 26:167–174

Thon G, Klar AJ (1992) The clr1 locus regulates the expression of the cryptic mating-type loci of fission yeast. Genetics 131:287–296

Trelogan SA, Martin SL (1995) Tightly regulated, developmentally specific expression of the first open reading frame from LINE-1 during mouse embryogenesis. Proc Natl Acad Sci USA 92:1520–1524

Van Gent DC, Hiom K, Paull TT, Gellert M (1997) Stimulation of V(D)J cleavage by high mobility group proteins. EMBO J 16:2665–2670

Van Luenen HG, Plasterk RH (1994) Target site choice of the related transposable elements Tc1 and Tc3 of Caenorhabditis elegans. Nucleic Acids Res 22:262–269

Vanegas M, Llano M, Delgado S, Thompson D, Peretz M, Poeschla E (2005) Identification of the LEDGF/p75 HIV-1 integrase-interaction domain and NLS reveals NLS-independent chromatin tethering. J Cell Sci 118:1733–1743

Vigdal TJ, Kaufman CD, Izsvak Z, Voytas DF, Ivics Z (2002) Common physical properties of DNA affecting target site selection of Sleeping Beauty and other Tc1/mariner transposable elements. J Mol Biol 323:441–452

Walisko O, Ivics Z (2006) Interference with cell cycle progression by parasitic genetic elements: Sleeping Beauty joins the club. Cell Cycle 5:1275–1280

Walisko O, Izsvak Z, Szabo K, Kaufman CD, Herold S, Ivics Z (2006) Sleeping Beauty transposase modulates cell-cycle progression through interaction with Miz-1. Proc Natl Acad Sci USA 103:4062–4067

Walisko O, Schorn A, Rolfs F, Devaraj A, Miskey C, Izsvak Z, Ivics Z (2008) Transcriptional activities of the Sleeping Beauty transposon and shielding its genetic cargo with insulators. Mol Ther 16:359–369

Walsh CP, Chaillet JR, Bestor TH (1998) Transcription of IAP endogenous retroviruses is constrained by cytosine methylation. Nat Genet 20:116–117

Winckler T, Szafranski K, Glockner G (2005) Transfer RNA gene-targeted integration: an adaptation of retrotransposable elements to survive in the compact Dictyostelium discoideum genome. Cytogenet Genome Res 110:288–298

Wu X, Li Y, Crise B, Burgess SM (2003) Transcription start regions in the human genome are favored targets for MLV integration. Science 300:1749–1751

Xie W, Gai X, Zhu Y, Zappulla DC, Sternglanz R, Voytas DF (2001) Targeting of the yeast Ty5 retrotransposon to silent chromatin is mediated by interactions between integrase and Sir4p. Mol Cell Biol 21:6606–6614

Yang N, Zhang L, Zhang Y, Kazazian HH Jr (2003) An important role for RUNX3 in human L1 transcription and retrotransposition. Nucleic Acids Res 31:4929–4940

Yant SR, Kay MA (2003) Nonhomologous-end-joining factors regulate DNA repair fidelity during Sleeping Beauty element transposition in mammalian cells. Mol Cell Biol 23:8505–8518

Yant SR, Meuse L, Chiu W, Ivics Z, Izsvak Z, Kay MA (2000) Somatic integration and long-term transgene expression in normal and haemophilic mice using a DNA transposon system. Nat Genet 25:35–41

Yant SR, Wu X, Huang Y, Garrison B, Burgess SM, Kay MA (2005) High-resolution genome-wide mapping of transposon integration in mammals. Mol Cell Biol 25:2085–2094

Yergeau DA, Mead PE (2007) Manipulating the Xenopus genome with transposable elements. Genome Biol 8(Suppl 1):S11

Yieh L, Kassavetis G, Geiduschek EP, Sandmeyer SB (2000) The Brf and TATA-binding protein subunits of the RNA polymerase III transcription factor IIIB mediate position-specific integration of the gypsy-like element, Ty3. J Biol Chem 275:29800–29807

Yieh L, Hatzis H, Kassavetis G, Sandmeyer SB (2002) Mutational analysis of the transcription factor IIIB-DNA target of Ty3 retroelement integration. J Biol Chem 277:25920–25928

Yin JC, Krebs MP, Reznikoff WS (1988) Effect of dam methylation on Tn5 transposition. J Mol Biol 199:35–45

Yoder JA, Walsh CP, Bestor TH (1997) Cytosine methylation and the ecology of intragenomic parasites. Trends Genet 13:335–340

Yu F, Zingler N, Schumann G, Stratling WH (2001) Methyl-CpG-binding protein 2 represses LINE-1 expression and retrotransposition but not Alu transcription. Nucleic Acids Res 29:4493–4501

Yusa K, Takeda J, Horie K (2004) Enhancement of Sleeping Beauty transposition by CpG methylation: possible role of heterochromatin formation. Mol Cell Biol 24:4004–4018

Zayed H, Izsvák Z, Khare D, Heinemann U, Ivics Z (2003) The DNA-bending protein HMGB1 is a cellular cofactor of Sleeping Beauty transposition. Nucleic Acids Res 31:2313–2322

Zhang L, Dawson A, Finnegan DJ (2001) DNA-binding activity and subunit interaction of the mariner transposase. Nucleic Acids Res 29:3566–3575

Zhao Y, Cao J, O'Gorman MR, Yu M, Yogev R (1996) Effect of human immunodeficiency virus type 1 protein R (vpr) gene expression on basic cellular function of fission yeast Schizosaccharomyces pombe. J Virol 70:5821–5826

Zou S, Ke N, Kim JM, Voytas DF (1996) The Saccharomyces retrotransposon Ty5 integrates preferentially into regions of silent chromatin at the telomeres and mating loci. Genes Dev 10:634–645

Genome Dyn Stab (4)
D.-H. Lankenau, J.-N. Volff: Transposons and the Dynamic Genome
DOI 10.1007/7050_2008_043/Published online: 14 January 2009
© Springer-Verlag Berlin Heidelberg 2009

Interactions of Transposons with the Cellular DNA Repair Machinery

Zsuzsanna Izsvák[1,2] (✉) · Yongming Wang[1] · Zoltán Ivics[1]

[1]Max Delbrück Center for Molecular Medicine, Robert Rössle Str. 10, 13092 Berlin, Germany
zizsvak@mdc-berlin.de

[2]Institute of Biochemistry, Biological Research Center of the Hungarian Academy of Sciences, Temesvári krt. 62, 6701 Szeged, Hungary

Abstract Transposons are ubiquitous components of both prokaryotic and eukaryotic genomes. Transposons are mobile genetic elements whose movement inherently involves the generation of various types of DNA damage in the genome. Since transposons generally do not encode DNA repair functions, the damage is left to be repaired by the host cell. Transposon-inflicted DNA damage ranges from single-base mismatches to the most severe double-strand breaks, and alert specific DNA repair pathways. The interaction between transposons and the host repair machinery is not necessarily a passive process. The host evolves inhibitory strategies against transposition, whereas transposons try to quench damage signaling and escape from inhibitory regulations. Moreover, certain transposons are able to "sense" activation of DNA damage signaling pathways that may eventually trigger transposition. The DNA repair machinery can even be piggybacked by transposons to support transposon amplification. This review focuses on the complex interactions of transposable elements with the DNA repair machinery. Additionally, the impact of mobile elements on genome stability is discussed.

Abbreviations

BER	Base excision repair
DSB	Double-strand break
HDR	Homology-dependent repair
IN	Integrase
L1	*Line1*
MLV	Murine Leukemia Virus
MMR	Mismatch repair
NER	Nucleotide excision repair
NHEJ	Nonhomologous end-joining
PIC	Preintegration complex
PRR	Postreplication repair
RIP	Repeat-induced point mutation
RSS	Recombination signal sequence
RSV	Rous Sarcoma Virus
SDSA	Synthesis-dependent strand annealing
SSA	Single-strand annealing
TPRT	Target-site primed reverse transcription
TSD	Target site duplication

1
Introduction

Transposable elements are components of both prokaryotic and eukaryotic genomes, and are best described as molecular parasites. Transposons encode the protein machinery (a recombinase) and occasionally additional proteins that are necessary for the transposition reaction. Transposition inherently involves the generation of DNA damage in the genome and, in general, transposons leave the damaged DNA behind to be repaired by the host cell. Transposition can be damaging to gene function, but there are also examples of transposons contributing cellular functions that are useful for the host (Feschotte and Pritham 2007; McClintock 1984). Thus, transposable elements and the host have coevolved in a way that permits propagation of the transposon, but minimizes damage to the host. The evolutionary success of transposable elements is powerfully underscored by the finding that, for example, about 45% of the human genome is transposon-derived.[1] This review focuses on the repair of DNA gaps/lesions inflicted by transposons and the interactions of transposons with the DNA repair machinery during transposition.

2
The Types of DNA Damage Produced by Transposons

Transposable elements can transpose via a DNA (DNA transposons) or an RNA intermediate (retroelements). DNA transposons physically separate from the donor site and reintegrate somewhere else in the genome (cut&paste transposition), while retroelements produce a copy of their genome that integrates, but they never leave the donor site (copy&paste transposition). A transposition-like process in vertebrates that has been extensively studied is immunoglobulin gene rearrangement, also termed V(D)J recombination (Gellert 2002) (Fig. 1). DNA transposons can generate DNA damage at the sites of excision and integration. Transposon excision generates double-strand breaks (DSBs). In general, excision sites generated by transposition are specific to the transposon and seem to be different from DSBs generated by chemicals, radiation, or nucleases. Some transposons generate DSBs that are blunt ended (Tn5), while others generate a staggered cut, producing single-stranded DNA of different lengths (*P element*—17 bp, *Sleeping Beauty*—3 bp). The excision site can have a strong secondary structure, such as a hairpin (V(D)J recombination, *hAT* elements) or a heteroduplex (*Sleeping Beauty*, *Minos*). Additionally, the cleavage of the two strands of the DNA

---

[1] International Human Genome Sequencing Consortium (2001) Initial sequencing and analysis of the human genome. Nature 409:860–921

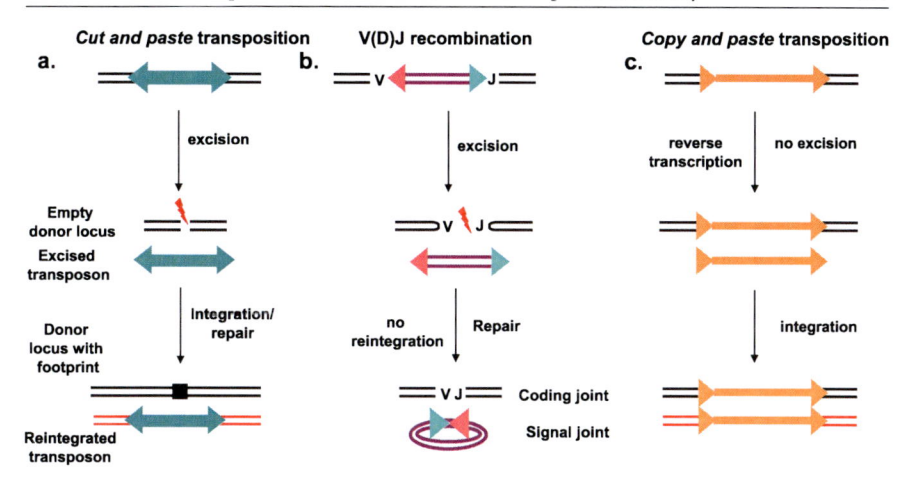

Fig. 1 The major types of recombination mechanisms in vertebrates. **a** Cut&paste transposition. DNA transposons physically separate from the donor site and reintegrate somewhere else in the genome. **b** V(D)J recombination is a transposition-like process in vertebrates. The excision step is similar to cut&paste transposition; however, the reintegration step is inhibited in vivo. **c** Copy&paste retrotransposition. Retroelements produce a copy from their genome that integrates, but they do not leave the donor site

is not necessarily synchronized, as was observed in *mariner* transposition (Dawson and Finnegan 2003). The integration intermediates are usually believed to consist of single-stranded DNA gaps flanking the newly integrated element (Yoder and Bushman 2000). Nevertheless, these intermediates of L1 retrotransposons (Gasior et al. 2006) and retroviruses (Skalka and Katz 2005) were proposed to resemble and be handled by the cell as DSBs. In addition, during DNA replication, single-stranded gaps have the capacity to either block the replication process, or to be converted into DSBs when a replication fork encounters such a lesion (Rothstein et al. 2000). In summary, structurally very different lesions can be generated by transposable elements that can trigger different DNA damage signaling mechanisms.

3
Cellular Processes Potentially Involved in Signaling and Repairing Transposition Intermediates

To properly protect the genome from the consequences of transposition, all types of DNA structural changes described above, including single-stranded nicks, heteroduplexes, gaps, DSBs, and blocks at replication forks, must be detected and repaired. The cellular response to DNA damage includes cell-cycle arrest, induction of DNA repair, or in the case of irreparable DNA damage, induction of apoptosis (see recent review by Riley et al. 2008). If the cell decides to repair the DNA damage, the pathway choice is differentially regulated depending on the type of lesion, the presence or absence of different host factors, the cell type, the degree of chromatin condensation, the developmental stage, and the cell cycle phase during which transposition occurs. According to our recent knowledge, at least five different molecular complexes sense and signal various types of DNA damage (reviewed in Harper and Elledge 2007). These complexes trigger the activation of diverse kinases, including the phosphoinositide-3-kinase-related protein kinase (PIKK) family members ATM, ATR, and DNA-PKcs, which are the primary transducers in the signaling cascade. While ATR activation is associated with single-stranded DNA and stalled DNA replication forks, ATM and DNA-PKcs respond mainly to DNA DSBs (Durocher and Jackson 2001). Mediators are proteins that act directly downstream of the PIKK kinases. Mediators act both as recruiters of additional substrates and as scaffolds upon which complexes assemble. These factors include Mdc1, 53BP1, the MRN complex, Claspin, Brit1/Mcph1, Brca1, and H2AX (reviewed in Harper and Elledge 2007).

Double-strand break repair (DSB repair) plays a key role in the maintenance of genome integrity (reviewed in Burma et al. 2006; Daley et al. 2005; Lees-Miller and Meek 2003; Rouse and Jackson 2002; Shrivastav et al. 2008) (Fig. 2a). In addition to DSBs generated by transposition, DSBs are also produced by exogenous genotoxic agents, such as ionizing radiation or chemicals, and by cellular processes, such as DNA replication, DNA breakage during meiosis (Keeney and Neale 2006), and V(D)J recombination (Franco et al. 2006). DSBs can be repaired by two major repair pathways (Baumann and West 1998; Haber 2000): nonhomologous end-joining (NHEJ) and homology-dependent repair (HDR). Compared to NHEJ, HDR is a more accurate pathway since it uses homologous sequence information to repair DSBs (reviewed in Krogh and Symington 2004; Paques and Haber 1999). However, single-strand annealing (SSA, a type of HDR) is a homology-dependent but error prone pathway, joining complementary single-stranded DNA after resection at a DSB. Thus, in the SSA process, the repair always results in deletion between the two homologous sequences flanking the DSB (Weinstock et al. 2006). Rad51, a structural and functional homolog of the *Escherichia coli* RecA protein, plays a key role in HDR (Sonoda et al. 2001). In mammalian cells, in addition to Rad51, repair factors including Rad52, Rad54, the Rad51 paralogs, and furthermore the breast cancer susceptibility proteins Brca1 and Brca2 all cooperate in HDR (Fig. 2a). During synthesis-dependent HDR, the resected ends of the DSB invade undamaged sequences from a homologous template located on a sister chromatid, a homologous chromosome, or at an ectopic location. HDR intermediates can be resolved by "classical" homologous recombination resulting in crossover or by synthesis-dependent strand annealing (SDSA) resulting in gene conversion. The SDSA process does not produce crossover products and may involve multiple rounds of strand invasion (compare Fig. 9, D.-H. Lankenau, vol. 1 this SERIES).

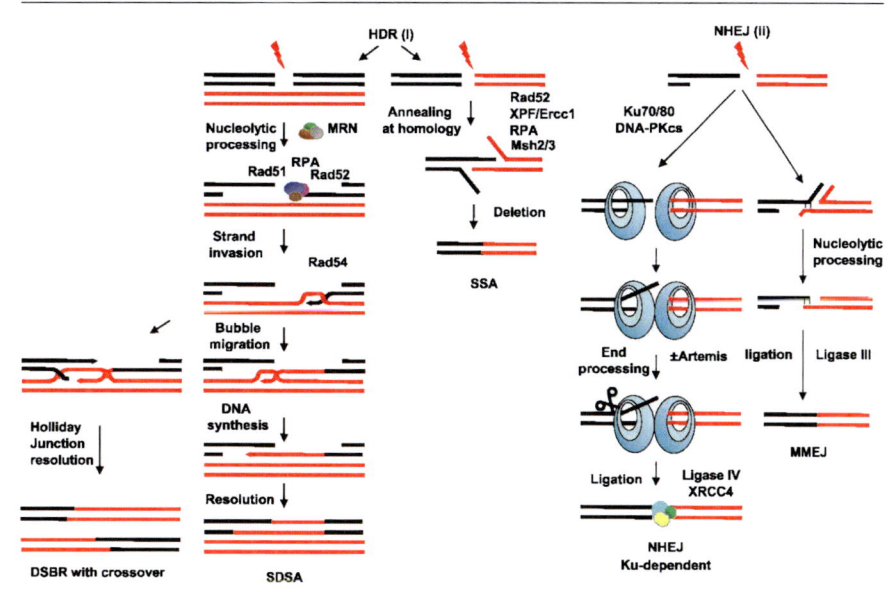

Fig. 2 Double-strand break repair (DSB repair) in mammalian cells. (i) Homology-dependent repair (HDR); (ii) nonhomologous end-joining (NHEJ). The HDR pathway can be subdivided into at least three subpathways. In the single-strand annealing (SSA) process, the homologous sequences, originally flanking the DSB, are annealed, resulting in the deletion of the DNA sequence between the repeats. During synthesis-dependent HDR, DNA ends are first processed in order to create single-stranded overhangs by the MRE11/RAD50/NBS1 (MRN) complex. Rad51 promotes a nucleoprotein filament assembly on the single-stranded ends of the broken DNA. Brca2 is assumed to control the process of the Rad51 nucleoprotein filament assembly. Rad54, a dsDNA-dependent motor protein, acts in stabilizing the Rad51 filaments on single-stranded DNA in an ATP-dependent fashion (Mazin et al. 2003), but was proposed to have functions at all three stages of HDR: presynapsis, synapsis, and postsynapsis (Ehmsen and Heyer (2008), this SERIES). During synthesis-dependent strand annealing (SDSA), the invading strand is disengaged concomitantly with DNA synthesis within a small, moving replication bubble, displacing the newly synthesized strand. The replication bubble is maintained until the second end could anneal with this new strand. Then, a second primer extension would be initiated to complete DSB repair without crossing over (Formosa and Alberts 1986; Engels 1994). In the "classical HDR", both ends of the DSBs are engaged, leading to Holliday junction formation. The junction can be processed into a noncrossover (not shown) or a crossover product. During NHEJ, broken ends are ligated together, often following limited resection to expose sites of microhomology, which are utilized in the ligation. The two subpathways of NHEJ are a dominant, Ku-dependent (or "classical") NHEJ and the Ku-independent (or "alternative") NHEJ. The Ku heterodimer (composed of Ku70 and Ku80) binds to DNA ends, and facilitates DSB repair by recruiting DNA-PKcs and additional factors such as the Xrcc4/DNA ligase IV complex to the site of damage. DNA-PK can activate the nuclease *Artemis* that helps repair a subset of DSBs, including hairpins formed during V(D)J recombination (Jeggo and O'Neill 2002). The "alternative", ligase III-mediated NHEJ is associated with microhomology, hence its name microhomology-mediated end-joining (MMEJ) (Wang et al. 2006)

In contrast to HDR, NHEJ is not dependent on a homologous template. The NHEJ process is error prone and after some processing (limited recession, hairpin resolution) simply rejoins the two ends of the broken DNA (see also Pfeiffer et al., this SERIES). Two different subpathways have been described for NHEJ: a dominant, Ku-dependent (or "classical") NHEJ and the Ku-independent (or "alternative") NHEJ. In mammals, the breaks are prepared for NHEJ by limited end-processing by the MRN complex. The main factors that mediate the "classical" NHEJ in mammalian cells are a complex of the DNA-dependent protein kinase (DNA-PK), DNA ligase IV, and Xrcc4. DNA-PK consists of a catalytic subunit (DNA-PKcs) and a DNA-binding subunit termed Ku. DNA-PK can activate the nuclease *Artemis* that helps to repair a subset of DSBs, including hairpins formed during V(D)J recombination (Jeggo and O'Neill 2002). The "alternative", ligase III-mediated NHEJ is more error prone than the "classical" NHEJ and functions as a backup pathway. It often utilizes limited resection to expose sites of microhomology (Wang et al. 2006). While in yeast the dominant pathway to repair DSBs is HDR, in higher eukaryotes NHEJ is more prominent, and HDR utilization is strictly controlled (Moynahan and Jasin 1997). In vertebrate cells, HDR and NHEJ have overlapping roles in maintaining chromosomal integrity (Takata et al. 1998), and represent alternative solutions to repair the same DSB (Baumann and West 1998; Haber 1992; Lankenau and Gloor 1998). Moreover, DSBs can be repaired by a combination of HDR and NHEJ (Adams et al. 2003; Couedel et al. 2004; Robert and Bessereau 2007).

The mammalian DNA mismatch repair (MMR) factors belong to a family of genes that comprise several homologs of the *E. coli* mutS and mutL genes (Weia et al. 2002; reviewed in Li 2008) (Fig. 3a). In eukaryotes, the initiation of the repair process requires subsets of three different MutS homologs: MSH2, MSH3, and MSH6. The MSH2/MSH6 complex is required for the repair of base–base mispairs as well as single-base insertion/deletion loops (IDLs). Most DNA lesions are removed prior to DNA replication by the nucleotide excision repair (NER) (Fig. 3b) and base excision repair (BER) pathways (reviewed in Shuck et al. 2008; Hegde et al. 2008). The NER pathway removes DNA structures caused by UV irradiation and chemical mutagens. When lesion-specific pathways (MMR, NER, BER) are unable to repair DNA lesions prior to the initiation of S phase, replication forks accumulate at sites of DNA damage, and can activate the postreplication repair (PRR) pathway (Fig. 3c). PRR allows DNA synthesis to reinitiate without removal of the offending lesion (Broomfield et al. 2001). Two modes of PRR exist: an error-prone mode during which the sites of DNA damage are bypassed without removing the errors, and an error-free mode that is believed to proceed by a template switching event. The members of the RAD6 pathway, including RAD6, RAD18, RAD5, MMS2, and UBC13, have been postulated to be involved in PRR (Broomfield et al. 2001). PCNA plays essential roles in both replication and several DNA repair pathways, such as MMR, long-patch BER,

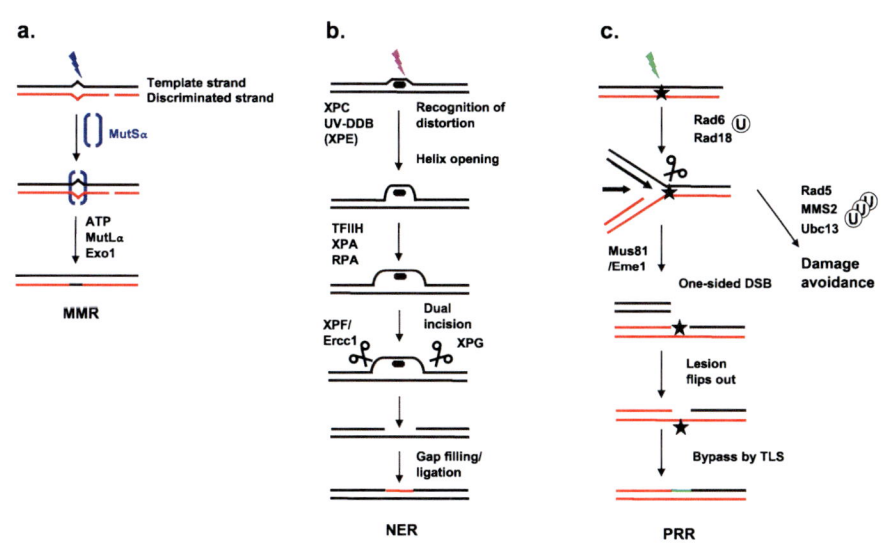

Fig. 3 Additional cellular DNA repair processes potentially involved in repairing transposition intermediates. **a** The mammalian DNA mismatch repair (MMR) performs recognition and editing functions. MMR corrects single-base mismatches or short stretches (2–4 bps) of mispairing nucleotides (insertions and deletions). In mammals, the initiation of the repair process requires subsets of different MutS homologs. The MutSα complex consists of MSH2/MSH6 (1 bp) or MSH2/MSH3 (2–4 bps). The discriminated strand is marked. In the presence of ATP, MutLα is recruited to the MutSα complex and together they perform incisions. **b** The nucleotide excision repair (NER) pathway removes DNA structures caused by UV irradiation and chemical mutagens. The repair of DNA damage begins with a damage recognition step and assembly of a preincision complex, followed by excision of the damaged strand and gap-filling DNA synthesis. The six core factors that have been implicated in the damage recognition and dual incision steps of global genomic-NER (GG-NER) are the XPC-RAD23B complex, transcription factor IIH (TFIIH), XPA, replication protein A (RPA), XPG, and XPF-ERCC1. **c** When lesion blocks the replication fork, postreplication repair (PRR) might be initiated. First, the Rad6-Rad18 complex is recruited to the site of DNA damage. Two modes of PRR exist: an error-prone mode during which the sites of DNA damage are bypassed without removing the errors (translesion synthesis, TLS), and an error-free mode that is believed to proceed by a template switching, or recombinational repair event (damage avoidance). Ubiquitination of PCNA is dependent on RAD6 and RAD18 (Hoege et al. 2002; Stelter and Ulrich 2003). For the damage avoidance pathway PCNA is further ubiquitinated by the complex MMS2-Ubc13-Rad5

and NER (Hubscher et al. 2002; Paunesku et al. 2001). It has been shown that PCNA can initiate and channel PRR to distinct pathways, orchestrated by differential posttranslational modification of PCNA by ubiquitin or SUMO (Hoege et al. 2002).

The key function of the RecQ helicases is proposed to be the processing of recombination intermediates formed during DNA replication (reviewed in Hickson 2003). Bloom syndrome, a deficiency in the RecQ helicase BLM, results in hypersensitivity to DNA damaging agents. BLM was hypothesized to suppress mitotic crossovers by restoring stalled replication forks in an error-free manner. Due to its branch migration activity, BLM is able to dissolve double Holliday junctions, in concert with TOPOIIIa and BLAP75 (Raynard et al. 2006), and reinitiate replication forks by preventing a crossover outcome of recombination events (Wu and Hickson 2003).

4
The Main Classes of Transposons

4.1
Cut&Paste DNA Transposons, V(D)J Recombination

A typical member of this group of mobile elements is the *Sleeping Beauty* transposon (Walisko et al., this BOOK). *Sleeping Beauty* is a member of the *Tc1* superfamily of elements that are probably the most widespread DNA transposons in nature (Plasterk et al. 1999). *Sleeping Beauty* consists of a single gene encoding the transposase polypeptide, the enzymatic factor of transposition, which is flanked by terminal, inverted repeats (IRs) containing binding sites for the transposase (Ivics et al. 1997). *Sleeping Beauty* is a valuable experimental system for probing transposon–host interactions in vertebrate organisms (Izsvák et al. 2000) (Fig. 4).

4.1.1
Structure of the Transposition Intermediate

DSB Repair Following Transposon Excision
The repair of the excision site is a DSB repair event, fine-tuned to gaps generated by transposons. *Tc1* transposases, retroelement integrases, and the RAG1 of V(D)J recombinase all contain an evolutionarily and functionally conserved amino acid triad, the DDE motif, within the catalytic domain that is responsible for both the excision and integration reactions. These recombinases catalyze a remarkably similar overall chemistry of DNA recombination (Craig 1995). During transposition of many bacterial transposons, *hAT* superfamily elements, and V(D)J recombination, a liberated DNA end (that results from nicking by the recombinase) is directed to attack the opposite, complementary DNA strand, thereby forming a hairpin intermediate that separates the transposon from the donor site. Alternatively, both strands are cut separately by the transposase, and no hairpin is formed in *mariner* (Dawson and Finnegan 2003) or *Sleeping Beauty* (Izsvák et al. 2004) transposition (Fig. 5a). In somatic cells, cut&paste-type excision sites are preferentially repaired by NHEJ (Izsvák et al. 2004). In addition to the basic factors of the NHEJ pathway, *Artemis* is required to repair the hairpin-structured excision

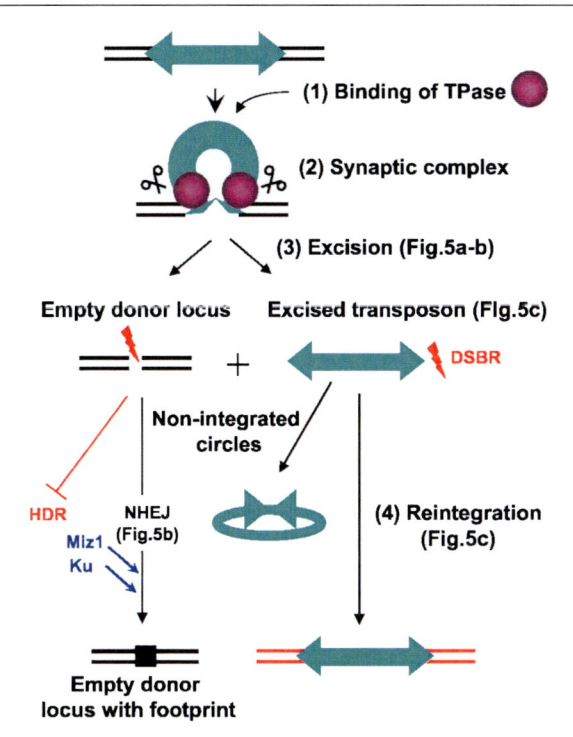

Fig. 4 Cut&paste transposition. In somatic cells, *Sleeping Beauty* transposes through a conservative, cut&paste mechanism, during which the transposable element is excised from its original location by the transposase, and is integrated into a new location. The transposition process can arbitrarily be divided into the following major steps: (1) binding of the transposase to its sites within the transposon IRs; (2) formation of a synaptic complex in which the two ends of the elements are paired and held together by transposase subunits; (3) excision from the donor site leaving behind a DSB; and (4) reintegration at a target site. Both the excision site and the excised transposon trigger DSB repair. In somatic cells, the *Sleeping Beauty* transposase promotes NHEJ by interacting with cellular host factors of Ku and Miz1, while HDR is inhibited

site (Fig. 5b). If the transposon is excised by a staggered cut, DNA repair of the broken DNA ends by NHEJ generates transposon "footprints". Direct ligation of the three-base-long 3′ overhangs at the excision site after *Sleeping Beauty* excision would generate a one-base mismatch at the center. Similar heteroduplexes have been observed at *Minos* element excision sites in *Drosophila* (Arca et al. 1997), but the genes responsible for its resolution are not known. The expectation is that such a mismatch would primarily be repaired by MMR.

Extrachromosomal Transposon Molecules
It was observed that during transposition of *Tc1*, *Tc3*, *Minos*, and the *Ac* element, excised but not reintegrated molecules can be detected (Arca et al. 1997; Gorbunova and Levy 1997; Radice and Emmons 1993; van Luenen et al. 1994).

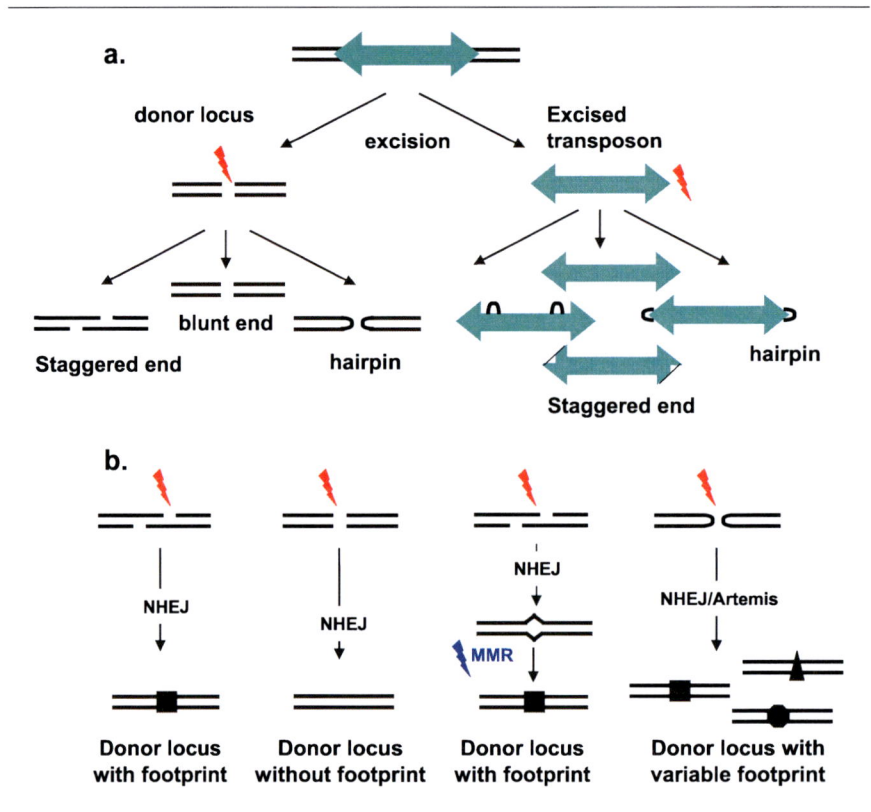

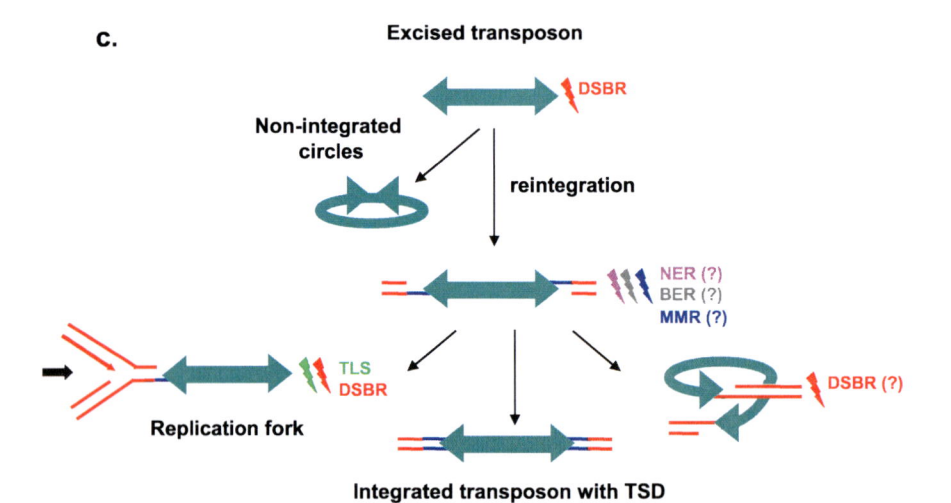

◀ **Fig. 5** Dissection of cut&paste transposition. **a** The structure of the excision site and the excised transposon. During transposition, a hairpin intermediate might be formed. Note: Although all the excision products are generated from a nicked intermediate, only the end products are shown. Depending on the primary cleavage step (3′ or 5′), the hairpin might be formed at the end of the transposon (Tn10) or at the excision site (hAT elements, V(D)J recombination) (reviewed in Turlan and Chandler 2000). Alternatively, no hairpin is formed. The cleavage by the transposon can be blunt or staggered. Similarly, the excised transposon might have a blunt, hairpin, or staggered end. Certain transposons have a strong secondary structure near their inverted repeats. **b** In somatic cells, excision sites are preferentially repaired by NHEJ. If the transposon is excised by a staggered cut, DNA repair of the broken DNA ends by NHEJ generates transposon "footprints". However, for certain transposons such as the *piggyBac* element, excision sites can be repaired without footprints (Mitra et al. 2008). At noncompatible gaps, in addition to DSB repair, MMR is involved in repairing the excision sites. Besides the basic factors of the NHEJ pathway, *Artemis* is required to repair the hairpin-structured excision site. The repair of hairpin-structured gaps generates diversity at the excision site, a basic feature utilized by V(D)J recombination, or hAT elements. **c** During transposition, excised but not reintegrated molecules can be detected. The excised molecules trigger repair signaling, and are neutralized by circularization. The joining reaction is assumed to produce an integration intermediate of single-stranded structure. DNA repair generates duplication of the target site (TSD) flanking the element. Very little is known about integration site repair. DNA pathways such as MMR, NER, or BER are the candidates involved. Postreplication repair (PRR) or RecQ helicases might be involved in avoiding replication arrest at the gapped transposition intermediates. Unrepaired gaps during replication become DSBs. It is also a possibility that integration intermediates trigger DSB repair

The extrachromosomal molecules resemble the signal molecules of V(D)J recombination. The excised molecules can trigger repair signaling and are neutralized by circularization (Fig. 5c).

Reintegration of the Transposon

The joining reaction is assumed to produce an integration intermediate flanked by single-stranded gaps. DNA repair at these gaps restores the terminal nucleotides of the inserted transposon, and generates a characteristic duplication of the target site (target site duplication, TSD) flanking the element (Fig. 5c). In comparison to excision site repair, almost nothing is known about integration site repair, although the host repair machinery is definitely involved. In the case of failure to repair the gaps at the integration sites immediately, cellular mechanisms that can attenuate the damage caused by transposition might have a crucial role in establishing stable host–transposon coexistence. PRR or RecQ helicases might play such a role. PRR is a strong candidate to confer damage tolerance that deals with the gapped transposition intermediates to avoid replication arrest. The fact that certain transposases were found to interact with members of PRR supports this assumption. PCNA was identified as an interactor of the *Pogo* transposase in *Drosophila* (Warbrick et al. 1998). In addition, a human relative of *Pogo*, named *Tigger*, encodes a transposase containing a conserved PCNA-binding domain (Warbrick et al. 1998). A similar motif is present in another *Pogo*-like element, *Lemi1* in *Arabidopsis*, suggesting that interaction with PCNA is a common feature in the *Pogo* family (Feschotte and Mouches 2000) (Table 1). It is also possible that the integration intermediate can be recognized as a DSB (Fig. 5c).

4.1.2
Regulation of DSB Repair at the Excision Site

Pathway choice for DSB repair following transposition might be determined by the transposon, the structure of the lesion, host-specific repair factors, the status of the cell cycle, the developmental stage of a particular cell, and whether transposition occurs in the soma or in the germline. The theory of passive competition between HDR and NHEJ at the DSBs has been modified, as "early acting" repair factors, associated with both NHEJ and HDR, such as BRCA1, H2AX, PARP-1, RAD18, DNA-PKcs, and ATM, were implicated to have an active control on pathway choice (reviewed in Shrivastav et al. 2008). Moreover, the phenomenon of mixed-pathway utilization suggests that one particular gap can be repaired by more than one repair pathway (Baumann and West 1998; Haber 1992).

V(D)J recombination can be viewed as a highly regulated transposition process (Fig. 6). An ancestral transposon, *Transib*, seems to have been adopted by the vertebrate immune system to generate immunoglobulin diversity (Kapitonov and Jurka 2005). Recombination is targeted by specific recombination signal sequences (RSSs), which are recognized and assembled by the V(D)J recombinase (transposase), RAG1/2, into a synaptic complex. V(D)J recombination can function as a real "cut&paste" type transposition reaction in vitro; however, the reintegration step is strictly inhibited in vivo. The cleavage reaction leaves a hairpin structure at the excision site (coding ends), with blunt ends on the excised transposon (signal ends). Compared to the transposon-inflicted DSBs, which can be repaired by both HDR and NHEJ pathways (Izsvák et al. 2004), RAG protein-inflicted gaps are repaired strictly by the NHEJ pathway (Gellert 2002) (Fig. 6). V(D)J recombination is restricted to the G1 phase of the cell cycle, where NHEJ is preferentially active. In addition, the postcleavage complex was proposed to guide DSBs to NHEJ, preventing the initiation of homologous recombination (Lee et al. 2004). Importantly, when classical NHEJ does not operate, not HDR but the alternative NHEJ functions as a background process (Corneo et al. 2007) (Fig. 6). The hairpin structure of the coding sequence of V(D)J recombination is also an important factor preventing SDSA initiation. Furthermore, the hairpin structure may direct transposition back to the donor site, playing a critical role in restricting transposition elsewhere in the genome (Posey et al. 2006). Similarly, the strong secondary structure of the excision site of

Table 1 Role of various factors/pathways in transposition. The roles of these factors involved in DNA repair, DNA damage signaling, and DNA replication, as well as their functions in DNA recombinational mechanisms, are indicated

Gene	Function	Role in DNA transposition	V(D)J recombination	Retro-transposition
H2AX	Sensing DSBs, regulation of pathway choice	No inf.	Required	HIV *L1*
ATM	DSB signaling, repair, regulation of pathway choice	*P element* *Sleeping Beauty*	Enforcing NHEJ, Not required for repair	Required for HIV or *L1* retromobility
ATR	Damage signaling during replication	No inf.		Required for HIV retromobility
Ku70/80 DNA-PKcs	NHEJ Damage signaling telomere maintenance, scaffold, regulation of pathway choice	*P element* *Sleeping Beauty* excision site repair	Coding and successful HIV signal joint formation	Required for integration; Ku is required for *Ty1* transposition; suppression of EN-independent *L1* integration
Artemis	Required to repair a subset (hairpin, etc.) of DSBs	Not required for *Sleeping Beauty* transposition	Coding joint formation	No inf.
Ligase IV/ XRCC4	NHEJ ligation	*P element* *Sleeping Beauty* excision site repair	Coding and signal joint formation	Required for successful HIV integration; suppression of EN-independent *L1* integration
Mre11/ Rad50/ Nbs1 (MRN)	HDR, NHEJ, DNA damage checkpoint activation	No inf.	Signal joint formation	Suppression of *Ty1* retromobility
RecQ (BLM, Sgs1)	Protecting from crossover outcome of recombination	*P element* excision site repair	Not required	Suppression of *Ty1* or MoMLV integration
PCNA	Replication, NER, long patch BER, MMR, PRR	*P element* *Pogo* *Tigger*	Not required	No inf.
Ssl2 (Rad25)	DNA helicase NER	No inf.	No inf.	Suppression of *Ty1* retromobility

Table 1 (continued)

Gene	Function	Role in DNA transposition	V(D)J recombination	Retro-transposition
Rad18	PRR, ubiquitin ligase, recruitment of Rad6; regulation of pathway choice	No inf.	No inf.	HIV, physical interaction; suppression of *Ty1* integration
Rad6	PRR	No inf.	No inf.	*Ty1* targeting
TfIIH	Transcription factor NER	No inf.		Suppression of *Ty1* or HIV integration
Fen1/ Rad27	Flap structure specific endonuclease; long patch BER	No inf.	Not required	Suppression of *Ty1* integration
MSH2/ MSH6 MLH1 PMS2	MMR	*Minos*, excision site repair	Not required	Suppression of HIV integration
Rad52	Homologous recombination	No inf.	Not required	Suppression of *Ty1* or HIV integration
Rad54	Homologous recombination	*P element* excision site repair	Not required	

certain transposons (*hAT* elements, including *Ac*), also has a role in pathway choice regulation (see later) (Fig. 7).

Besides structural determinants, species-specific host factors contribute as well to the regulation of pathway choice. For example, when *Ac* is transferred from maize to yeast (Weil and Kunze 2000), where HDR plays a more dominant role in repair of DSBs, DSB repair following *Ac* excision uses HDR more frequently. The inhibitory effect of the gap structure following *Ac* excision is still obvious, as the frequency of HDR utilization is still much lower compared to an endonuclease-induced DSB repair (Yu et al. 2004). The presence or absence of host factors influencing the overall chromatin structure has been postulated to affect DSB repair frequencies. As it was elegantly demonstrated in *Drosophila*, the genome-wide homology searches preceding DSB repair become more efficient when the overall chromatin structure is less constrained (Lankenau et al. 2000).

In mammalian cells, ATM and DNA-PKcs play a role in NHEJ, but they also influence HDR through a complex regulatory network. This regulation may involve several proteins of the HDR pathway that are phosphorylated

V(D)J recombination

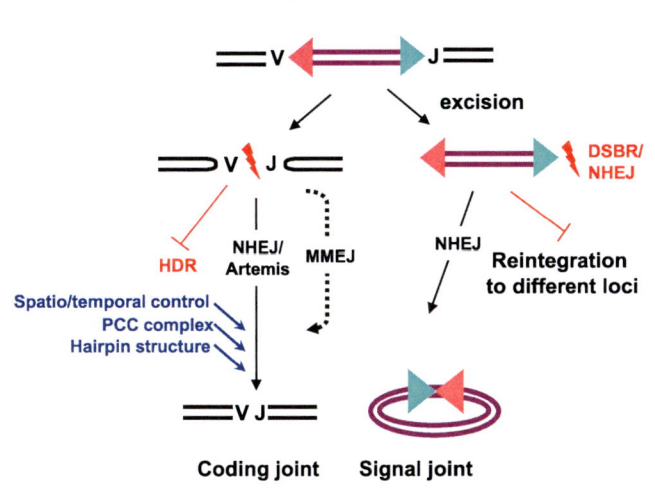

Fig. 6 V(D)J recombination. V(D)J recombination can function as a real cut&paste transposition in vitro; however, the reintegration step is inhibited in vivo. The cleavage reaction leaves a hairpin structure at the excision site (coding ends), and blunt ends on the excised transposon (signal ends). Gaps generated by V(D)J recombination are repaired strictly by the NHEJ pathway. There are multiple levels of regulation that prevent the cleaved transposon (signal joint molecule) from reintegration and enforce NHEJ, including spatio/temporal control of V(D)J recombination; pathway choice regulation by the postcleavage complex (PCC); the hairpin structure of the coding sequence. When classical NHEJ does not operate, the alternative NHEJ, MMEJ, functions as a background process

by DNA-PKcs and/or ATM. Importantly, the DNA-binding factor of DNA-PK, Ku, was reported to be associated with V(D)J recombination (Raval et al. 2008) and with SB transposition (Izsvák et al. 2004). Binding of Ku at the DSB is expected to influence the gap to be repaired preferentially by NHEJ (Fig. 7). Recently, DNA-PKcs was suggested to be a key regulator of pathway choice (Shrivastav et al. 2008). In *Sleeping Beauty* transposition, DNA-PKcs has a function that is independent from its kinase activity and was proposed to serve as a molecular scaffold (Izsvák et al. 2004). One can hypothesize that this scaffold function of DNA-PKcs is involved in regulating which repair factors would be recruited to initiate DSB repair. ATM is also on a list of candidates of pathway choice regulation. ATM was associated with *P element* transposition (Beall et al. 2002) as it binds the inverted repeats of the transposon (Beall and Rio 1996). The involvement of RecQ helicases in *P element* transposition is supported by the findings that the excision site repair requires BLM activity. In the absence of BLM, the excision sites of *P elements* have large deletions. Strikingly, this effect can be partially rescued by overexpressing Ku70 (one subunit of the Ku heterodimer) (Beall and Rio 1996), suggesting an interaction between BLM and the NHEJ pathway (Min et al. 2004).

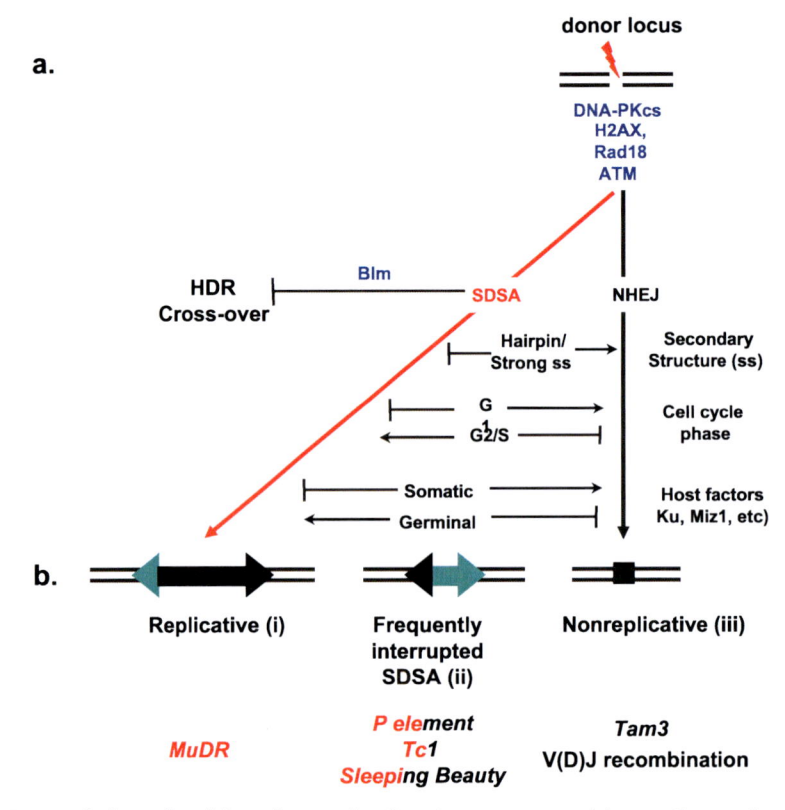

Fig. 7 Regulation of excision site repair of cut&paste transposition. **a** Factors involved in the pathway choice regulation following excision include strong secondary structure of the excision site, e.g., a hairpin structure aborts the SDSA process; cell cycle phase of the cell; cell type (somatic or germ). Host factors: (i) "early acting host factors", DNA-PKcs, H2AX, Rad18, ATM; (ii) host factors binding either specifically to the subterminal repeats of the transposon (ATM, *P element*) or to the (iii) transposase (Miz1/Ku70, *Sleeping Beauty*); BLM is assumed to suppress the crossover outcome of HDR resolution. **b** Correlation between successful SDSA utilization and transposon amplification. (i) *MuDR* successfully piggybacks the HDR in the germ cells to propagate its own genome. (ii) Frequently aborted SDSA produces a large number of nonautonomous copies with internal deletions. (iii) DSB repair is restricted to NHEJ; no SDSA is utilized

By using a repair reporter system in *Drosophila*, pathway choice was demonstrated to correlate with the developmental stage and the age of the organism. During early germline development, DSBs were more frequently repaired via SSA compared to NHEJ (Preston et al. 2006b). Regarding age, pathway choice was clearly shifting from the simpler end-joining processes to the more precise, template-dependent homologous mechanisms as the flies aged (Preston et al. 2006a).

Finally, DSBs would be differently treated in somatic and in germ cells. In mammalian somatic cells, the preferred way to repair a DSB is NHEJ (compare Pfeiffer et al., vol. 1 this SERIES). Although this pathway may lead to mutations at the repaired location, these are not passed on to the next generation. In contrast, in order to preserve accurate genetic information, the germline prefers to use the error-free HDR pathway. In addition, during meiosis, homologous recombination is deliberately induced by self-inflicted DSBs to facilitate exchanges of genetic information between homologous chromosomes. The DSBs in meiotic recombination induced by the evolutionarily conserved Spo11 protein are substantially different from any other DSBs generated in somatic cells, since Spo11 remains covalently bound to the 5' ends of the DSBs and the DSB is processed by specific endonucleolytic cleavage (see chapter by Keeney (2008), this SERIES). Transposons seem to have an active influence on pathway choice regulation that seems to be different in somatic and in germ cells. In somatic cells, DSB repair following *Sleeping Beauty* transposition is linked to the NHEJ pathway. The *Sleeping Beauty* transposase was proposed to be actively involved in manipulating the host repair by interacting with at least two cellular host factors, Ku and Miz1 (Izsvák et al. 2004; Walisko et al. 2006) (Fig. 4). As we discuss below, *MuDR* transposition is a role model of cell type-specific regulation of gap repair, using NHEJ in somatic cells and piggybacking HDR in germ cells to propagate its own genome (Hsia and Schnable 1996). In the next section, let's have a closer look at how transposons might utilize diverse cellular mechanisms, including DNA repair, in order to amplify their genome.

4.1.3
Strategies to Amplify the Genome of a Transposon

4.1.3.1
Transposon Amplification Utilizing DNA Repair

The model of cut&paste transposition is by definition nonreplicative: the element excises from one place, and inserts into another (Fig. 1). This raises the question as to how these elements can reach significant copy numbers in certain species (e.g., salmon contains on the order of 10 000 copies of *Tc1*-like transposons).

Gene Conversion, Synthesis-Dependent Strand Annealing (SDSA)
A possible model to explain how transposons amplify emerged during the evolution of DSB repair models. The classical DSB repair model is a "cut&paste"-type process that involves DNA synthesis, two migrating Holliday junctions, and extensive D-loop formation (Szostak et al. 1983). However, certain biological processes, such as mating-type switching in yeast, T4 bacteriophage recombination or gap repair following DNA transposition could not be subsumed under the explanatory umbrella of Szostak's classi-

cal model. The alternative, or gene conversion models describe DSB repair as a homology-dependent, yet nonreciprocal, recombination process.

The quintessence of gene conversion was already captured by the "cassette model" of mating-type switching in budding yeast by emphasizing the applicability of a unidirectional copy&paste (rather that cut&paste) DNA strand transfer (Egel et al. 1980; Hicks et al. 1977). The theoretical considerations of Resnick (1976), and the observations made during the T4 phage recombination process (Formosa and Alberts 1986) had a significant impact on our understanding of gap repair. In 1986 Formosa and Alberts took a brilliant in vitro approach, and demonstrated that the displacement of the newly synthesized strand can occur in a coupled system including a recombinase, DNA polymerase, and helicase enzymes, encoded by the *uvsX*, *gene43*, and *dda* genes, respectively (Formosa and Alberts 1986). Importantly, they observed that, instead of extensive D-loop formation, DNA synthesis occurs in a rather small replication bubble and hence proposed the "bubble migration model" (Formosa and Alberts 1986) (Fig. 2). SDSA emerged as a gene conversion model to explain *P element*-induced gap repair data in a metazoan system, *Drosophila* (Engels et al. 1990, 1994; Gloor and Lankenau 1998; Gloor et al. 1991; Lankenau 1995; Lankenau and Gloor 1998; Nassif et al. 1994). The models of bubble migration and SDSA are similar, except that SDSA emphasizes the strand-annealing feature over bubble migration. Importantly, in contrast to the "classical" DSB repair model (Szostak et al. 1983) that involves the formation of a double Holliday junction, the gene conversion models were most appropriate to describe the repair process of *P element* excision-induced DSBs, occurring almost always without an associated crossing-over (Nassif et al. 1994). In addition, the SDSA model was able to deal with the observation that some of the refilled, excised loci appeared to originate from more than one template, resembling independent template searches and template switches of each of the two ends of a DSB (Nassif et al. 1994). Again, such events were not anticipated by the classical DSB repair model but could be accounted for by SDSA. Finally, the observation that there were almost no changes at the template locus during excision site repair of *P element* transposition was fully compatible with an SDSA mechanism but inconsistent with the classical model, in which new DNA synthesis is found both at the template and the recipient DNA molecules.

Transposon amplification can be explained by gene conversion, i.e., the transposon sequence is filled back by a template-dependent repair process to the excision site (Lankenau 1995; Lankenau and Gloor 1998; Nassif et al. 1994). The gene conversion process is conserved from bacteria to multicellular organisms and by now it is considered as a replicative repair pathway that is frequently utilized to repair DSBs generated by various agents, including transposition (Arcangioli 2000; Formosa and Alberts 1986; Johnson and Jasin 2000; Lankenau 1995; Paques et al. 1998; Puchta 2005). Compared to the classical DSB repair, SDSA seems to be a safer process to deal with

DSBs associated with activities of repetitive elements (Lankenau et al. 2000). The classical DSB repair pathway would possess a high risk of generating crossover products that lead to scrambled chromosome rearrangements. Although SDSA might be a preferred pathway to repair transposition-induced DSBs, a recent report suggests that certain transposons, such as *MuDR* elements, can initiate meiotic crossovers (Yandeau-Nelson et al. 2005).

Rare Repair Synthesis at the Site of Transposon Excision

Strikingly, not all transposable elements utilize the SDSA strategy in a similar way to amplify their genome, and according to the examples below we can establish three distinct categories. First, as discussed earlier, V(D)J recombination is strictly linked to the NHEJ pathway of DSB repair (Jackson and Jeggo 1995), and HDR is not involved in the process. Transposable elements, possessing some degree of similarity to V(D)J recombination, provide good models on studying the regulation process. Deciphering the differential regulation of V(D)J recombination and transposition generated considerable attention (Roth 2003). The *hAT* family of transposons, present in plants, animals, and fungi, includes the transposable elements *Activator/Dissociation* (*Ac/Ds*) (maize), *Tam3* (snapdragon), *Hermes* (fly), *Tol2* (medaka fish), *Hobo* (*Drosophila*), and *Ascot-1* (fungus) (Calvi et al. 1991; Colot et al. 1998; Feldmar and Kunze 1991; Jasinskiene et al. 1998; Kawakami et al. 1998; Warren et al. 1994). Similarly to the coding joints of V(D)J recombination, the *hAT* family of transposons leave a hairpin intermediate at the excision site (Zhou et al. 2004). The strong secondary structure and additional repeats of the transposon itself seem to present a physical barrier to repair this type of lesion by HDR. The SDSA process is sensitive to the secondary structure of the DNA, and it frequently aborts at DNA structures, such as hairpins or repeats (Yamashita et al. 1999). The gap repair of de novo somatic or germinal *Tam3* excision is arrested shortly after initiation of the gene conversion process. Thus, all of the initiated gene conversion products contain only short stretches of freshly synthesized DNA (Yamashita et al. 1998, 1999). The gap repair halts within the hairpin structure regions, generating substantially deleted versions of the transposon. Other members of the *hAT* family, like the *Ac* element in maize, also have very rare repair synthesis at the site of excision (Dooner and Martinez-Ferez 1997; Scott et al. 1996; Yan et al. 1999).

A Frequently Aborted SDSA Process Can Still Contribute to Transposon Amplification

Similarly, NHEJ is primarily responsible for repairing the DNA damage inflicted by lesions in somatic cells following *Sleeping Beauty* (*Tc1* family) excision. Nevertheless, the dependence on NHEJ is not absolute. Especially when NHEJ is not available, HDR contributes to the repair process. In cells deficient for NHEJ, SDSA products are readily detectable (Izsvák et al. 2004). Both NHEJ factors and ATM seem to inhibit HDR pathways to access transposon excision sites. In the majority of cases the SDSA process is prematurely terminated, frequently finished by the "alternative" NHEJ pathway. It was concluded that the repair process is actively influenced by the transposon (Izsvák et al. 2004). Nevertheless, in contrast to V(D)J recombination or the *hAT* elements, both NHEJ and HDR play significant roles in the repair of DSBs generated by the transposons belonging to this group (Gloor et al. 2000; Kooistra et al. 1999). The *P element* transposase generates a gap flanked by 17-bp ssDNA at each side of the DSB. The average 30-bp-long footprint is the result of the NHEJ repair process and is not created by imprecise excision events of the transposase (O'Brochta et al. 1991). Significantly larger but still internally truncated *P element* remnants are interpreted as not the products of NHEJ but of incomplete SDSA activities. In premeiotic germ cells, SDSA is frequently utilized during *P element* (*Drosophila*) or *Tc1* (*Caenorhabditis elegans*) transposon excision site repair (Moerman et al. 1991; Plasterk and Groenen 1992) (Fig. 7b). In these cells, NHEJ accounts for less than one-third of the DSB repair products following *P element* excision (Johnson-Schlitz and Engels 1993). The gene conversion process was capable of copying several kilobases of DNA into the excision site (Nassif et al. 1994). The sister chromatid, containing a copy of the transposon in the same locus, can occasionally serve as a template for SDSA, which thus has the capacity to restore the element at the excision site (Johnson-Schlitz and Engels 1993; Nassif et al. 1994). When SDSA is utilized for DSB gap repair, the process is carried out in multiple cycles of strand invasion (McVey et al. 2004a). Thus, the gap repair process is sensitive to interruptions. In events when SDSA is initiated but the synthesis gets interrupted the repair might still be completed by SDSA, or by a different pathway (Izsvák et al. 2004; McVey et al. 2004b). The multiple SDSA cycles might also explain the structure of some *Tc1* elements in the nematode genome possessing a mosaic structure. It seems that during DSB repair after *Tc1* excision, the template can switch from the *Tc1* element on the sister chromatid or homologous chromosome to a *Tc1* copy elsewhere in the genome (Fischer et al. 2003). Template switching and capturing filler DNA sequences at DSBs generated by transposons has been documented in various organisms (Heslip et al. 1992; Lin and Waldman 2001; Nassif et al. 1994; Rubin and Levy 1997; Wessler et al. 1990; Yu and Gabriel 1999). In summary, despite efficient utilization of SDSA by this group of transposons, the synthesis process is frequently interrupted, giving rise to several internally deleted copies (Izsvák et al. 1995).

An Extremely Precise and Efficient SDSA Process Amplifies the *MuDR* Transposon

In the third group of transposons, represented by the maize *MuDR* transposon, SDSA utilization is extremely efficient and precise in germinal cells (Fig. 7b). During somatic development, both excision and integration of the *MuDR* transposon are readily detected, and the excision sites are repaired by the NHEJ pathway. However, it was proposed that *MuDR* insertions that occur late in germinal development are not associated with excisions at all (Alleman and Freeling 1986; Walbot and Warren 1988). It was recently clarified that in cells deficient for Rad51 (a key factor in HDR), partially deleted *MuDR* transposons can be recovered more frequently (Li et al. 2008). Thus, these deletion products can be explained by incomplete SDSA after *MuDR* excision. Extraordinarily, following excision of *MuDR* elements, the SDSA process replaces the excised element with an unexpected frequency and fidelity (Hsia and Schnable 1996). Since the germinal reversion frequency after *MuDR* excision is exceptionally low, the *MuDR* element can germinally duplicate at frequencies approaching 100% (Alleman and Freeling 1986; Walbot and Warren 1988). Therefore, it appears that SDSA is a very efficient and precise process capable of refilling the complete *MuDR* element into the excision site, resulting in the duplication of the transposon (Fig. 8a). Thus, the *MuDR* transposon efficiently piggybacks the cellular repair mechanism to replicate its own genome.

The Price to Pay ...
For the amplification of its genome by the SDSA pathway, the transposon has a price to pay. We still do not understand how the maize *MuDR* element manages to trick the host repair machinery into duplicating its genome with an extremely high frequency and faithfulness. However, in general, the cut&paste-type transposons in eukaryotes comprise a large number of nonautonomous copies with internal deletions (Doring and Starlinger 1986; Doring et al. 1984; McClintock 1950), and completely precise restoration of the autonomous copies is rare. In *P element* transposition, the internal deletions are estimated to occur about tenfold more frequently than precise gap repair (Delattre et al. 1995). As a consequence of frequent interruption of the SDSA process, active transposons produce numerous nonautonomous elements, diluting out autonomous copies from the genome (Engels et al. 1990). This process is definitely exposed to selective pressure by the gap repair mechanism of the host, and in most cases results in the complete loss of autonomous copies and fossilization of the transposon (Engels et al. 1990; Smit and Riggs 1996). Some transposons, like the *Tam3* element, appear to "fight" against the generation of defective copies via aborted SDSA gap repair with strong hairpin structures (Yamashita et al. 1999).

4.1.3.2
Jumping in Front of a Replication Fork as an Alternative Strategy to Amplify the Genome of the Transposon

Even if SDSA seems to be one potential way to contribute to transposon amplification, this process is not equally efficiently utilized at the excision sites of all transposable elements. Thus, alternative mechanisms should exist supporting transposon amplification (Fig. 8). In agreement with this assumption, the *Ac/Ds* elements in maize transpose shortly after replication from one of

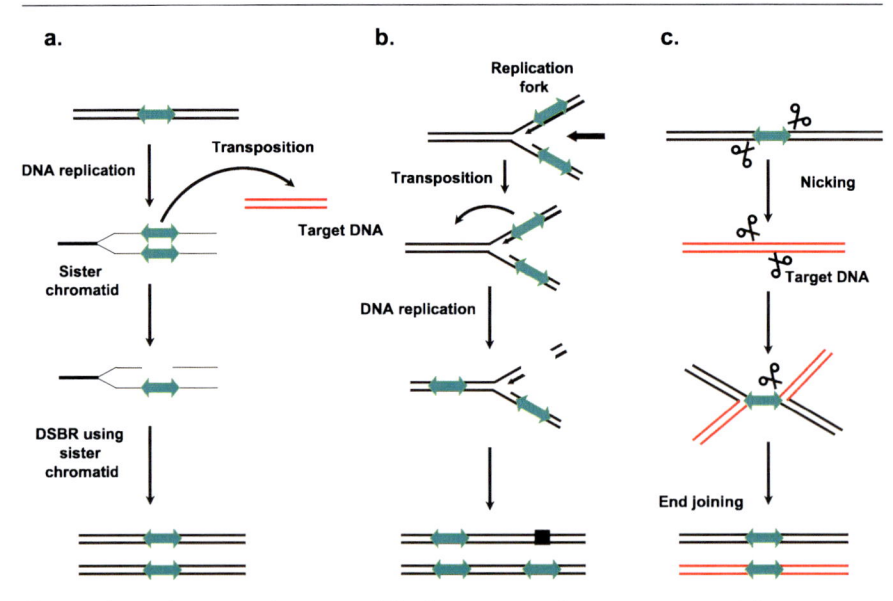

Fig. 8 Alternative strategies to amplify the genome of a transposon. **a** The transposition occurs in S phase during DNA replication. The DSB, generated at the donor site, is repaired by SDSA using the sister chromatid as a template. SDSA copies the entire transposon into the excision site. **b** Transposition occurs during DNA amplification. The transposon jumps in front of the replication fork. **c** The transposase produces nicks at both the donor and the recipient sites. The 3′ end of the transposon is ligated to the 5′ end of the target sequence. The transposon is duplicated after DNA replication

the two resulting chromatids ("chromatid selectivity"). A model was suggested that explains this phenomenon as a consequence of different *Ac* transposase binding to holo-, hemi-, and unmethylated transposon ends (Wang et al. 1996). It assumes that before replication the element is holomethylated and does not transpose because the transposase cannot bind to the transposon ends. Shortly after replication one of the two differentially hemimethylated daughter transposons becomes transposition-competent. Genetic evidence indicates that the majority of *Ac* transposition events in maize take place out of replicated chromosomal regions (Chen et al. 1987, 1992). The proposed mechanism requires transposon excision out of an already replicated locus after passage of a replication fork, but before completion of DNA synthesis. This strategy can lead to transposon amplification (Fig. 8b).

4.1.3.3
Switch from Nonreplicative to Replicative Mode of Transposition

Some DNA transposons, such as the *IS6* family, Tn3 (Mahillon and Chandler 1998), and bacteriophage *Mu* (reviewed in Lavoie and Chaconas 1996) can switch from a nonreplicative to replicative mode of transposition (Shapiro

1979). Both *Helitrons* and *Maveriks* are also thought to transpose in a replicative manner; although the mechanism is not yet understood, it is postulated to use a self-encoded DNA polymerase (Pritham et al. 2007). Strikingly, it was demonstrated that a single amino acid change in the bacterial Tn7 transposase causes it to switch from cut&paste transposition to replicative transposition; the amino acid change causes a block in 5'-DNA strand cleavage but not in 3'-cleavage or strand transfer (May and Craig 1996) (Fig. 8c). As a consequence, the DSB required for cut&paste transposition does not occur.

4.2
Copy&Paste Retroelements

Retroelements are genetic elements that, similarly to transposons, integrate themselves into the DNA of a host cell but, unlike classical transposable elements, do not leave the genomic site. Instead, retroelements propagate by replicating themselves through reverse transcription, and integrating new copies into new sites in genomes (Fig. 1c).

4.2.1
LTR Retrotransposons, Retroviruses

Long terminal repeat (LTR) retrotransposons and retroviruses belong to the same superfamily of retroelements. In addition to the characteristic LTR structure, their recombinase/integrase shares the conserved DDE motif in their catalytic domain, linking cut&paste transposons and LTR retroelements. Although the element does not excise from the donor locus during retrotransposition, free DNA ends with DSBs are still generated by reverse transcription, which produces a linear cDNA product (Li et al. 2001). The nonintegrated, linear cDNA is a potent substrate for the host DSB repair (Fig. 9). Similarly to cut&paste transposition, the integration process is thought to generate single-stranded integration intermediates. However, DSB repair might play a role also during retrotransposition integration (Skalka and Katz 2005) (Fig. 9).

The mobility of the *Ty1* LTR retrotransposon in yeast is restricted by a number of proteins that preserve the integrity of the genome. Approximately 30 genes (Maxwell and Curcio 2007) identified as repressors of *Ty1* transposition are "genome caretakers" and some of them, including Rad3, Ssl2, TFII, Rad52, Rad18, and Rad27/Fen1 (Bryk et al. 2001; Lee et al. 2000; Rattray et al. 2000; Sundararajan et al. 2003) are orthologs of mammalian retroviral restriction factors (Lau et al. 2004; Yoder et al. 2006). Some of these factors, namely the NER helicases TFIIH, Rad3, and Ssl2 (Lee et al. 2000), or Rad27/Fen1 involved in BER, are assumed to block *Ty1* cDNA accumulation at a step subsequent to reverse transcription (Lee et al. 2000; Sundararajan et al.

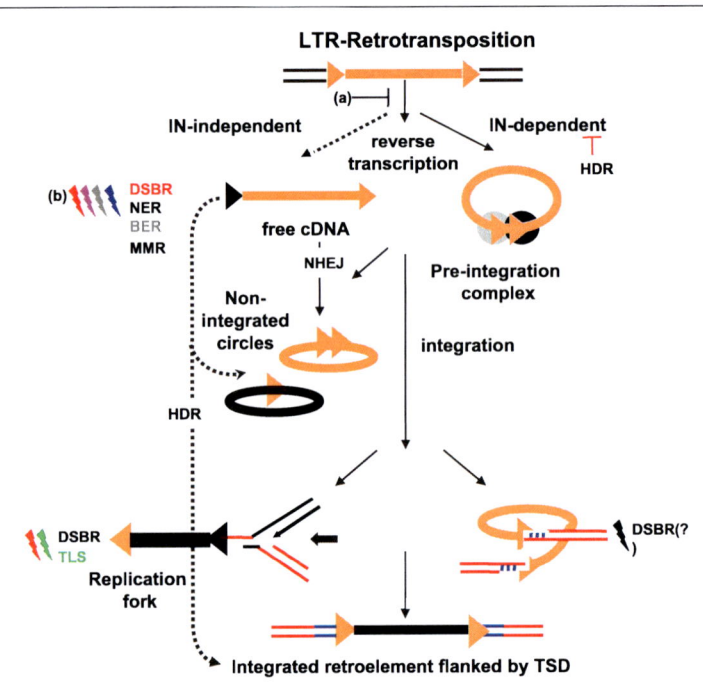

Fig. 9 LTR retrotransposition, retroviral integration. During retrotransposition the element does not excise from the donor locus. The nonintegrated, linear cDNA molecules, produced by reverse transcription, are potent substrates for DSB repair. The DNA damage pathway and/or the replication stress pathways can trigger retrotransposition by elevating reverse transcription. In yeast this process includes certain genome caretakers (a), Rad50, Mre11, Xrs2, Rad51, Rad52, and Rad55, primarily preventing the formation of DNA lesions. The genomic integration of a retrotransposon can occur by IN-independent or IN-dependent pathways. Rad52 and IN might compete for the cDNA of the retroposon. In the presence of IN, Rad52 inhibits retrotransposition; however, in the absence of competing IN, the cDNA is efficiently utilized for homologous recombination with one of the preexisting genomic elements. Several host factors influence the stability of the cDNA (b), Rad3, Ssl2, TFII, Rad52, Rad18, and Rad27/Fen1, etc. The NHEJ pathway of DSB repair is also involved in regulating the process of retrotransposition. DSB repair has been implicated in protecting cells from apoptosis caused by nonintegrated linear cDNA molecules. DSB repair is required to circularize nonintegrated viral cDNA (1- and 2-LTR circles). Alternatively, DSB repair might play a role during retrotransposition integration. Members of the PRR, Rad6, Rad18 have been postulated to play a role in regulating LTR retrotransposition

2003), interacting directly with *Ty1* cDNA (Sundararajan et al. 2003) (Fig. 9a). In the absence of these factors, the physical stability of broken DNA molecules and unincorporated *Ty1* cDNA is elevated, which leads to higher levels of *Ty1* retrotransposition. Similarly, it was demonstrated that the human TFIIH complex proteins of NER play a principal role in the degradation of retroviral cDNA (Yoder et al. 2006). MMR seems to protect genome integrity from retroviral integration as well. In MMR-deficient mammalian cells, retroviral integration was significantly elevated (Lin et al. 2001).

In addition to controlling cDNA stability, reverse transcription can also be affected. It was observed that the DNA damage pathway and/or the replication stress pathways can trigger cDNA accumulation and retrotransposition, by elevating *Ty1* reverse transcription (Curcio et al. 2007). These S-phase checkpoint pathways block cell cycle progression and initiate a transcriptional and posttranslational response to the DNA lesion. This model proposes that certain genome caretakers, including Rad50, Mrc11, Xrs2, Rad51, Rad52, and Rad55, are primarily preventing the formation of DNA lesions. In the absence of these factors, DNA lesions are not repaired properly, and activate DNA integrity checkpoint pathways that would stimulate *Ty1* cDNA synthesis (Fig. 9b). Perhaps the phenomenon that DNA damage response proteins are required for efficient retroviral integration might have a similar explanation (Daniel et al. 2003; Lau et al. 2005; Nunnari et al. 2005).

In general, the genomic integration of *Ty1* requires specific terminal nucleotides present in full-length *Ty1* cDNA. However, integrase (IN) mutant *Ty1*s are still able to efficiently integrate into the genome (Fig. 9). This IN-independent process is not a true retrotransposition process, and requires RNA intermediate, reverse transcriptase (RT), and protease (Sharon et al. 1994). The Rad52-mediated HDR inhibits "true" *Ty1* transposition (Rattray et al. 2000), but is required for the IN-independent pathway. One explanation can be that Rad52 and IN are competing for the cDNA and in the absence of competing IN, *Ty1* cDNA is efficiently utilized for homologous recombination with one of the preexisting genomic *Ty1* elements (Radford et al. 2004). This competition model between Rad52 and the IN is further supported by the observation that the mammalian Rad52 also has a negative effect on retroviral retrotransposition (Lau et al. 2004).

The NHEJ pathway of DSB repair is also involved in regulating the process of retrotransposition. Ku has been shown to be associated with *Ty1* (Downs and Jackson 1999) as well as with retroviral particles (Li et al. 2001). In the absence of Ku, *Ty1* as well as retroviral transposition frequency is decreased. The DNA repair of retroviral integration intermediates was proposed to depend on components of the Ku-dependent NHEJ pathway (DNA-PKcs, XRCC4, LigaseIV) and DNA damage-sensing pathways (ATR/ATM (Lau et al. 2005), gamma-H2AX (reviewed in Skalka and Katz 2005)). Retroviral integration-generated strand breaks might resemble DSBs, and were proposed to activate NHEJ. Alternatively, the NHEJ DNA repair pathway has been implicated in protecting cells from retrovirus-induced apoptosis caused by nonintegrated linear viral DNA. NHEJ activity was found to be required for an end-ligation reaction that circularizes a portion of the nonintegrated viral cDNA in infected cells (2-LTR circles). In contrast, a different form of circularized molecules, 1-LTR circles, is produced in part by homologous recombination (Kilzer et al. 2003).

Additionally, PRR has been postulated to play a role in regulating LTR retrotransposition. The HIV integrase was shown to physically interact with RAD18 (Mulder et al. 2002), and RAD6 was reported to influence both the frequency and target site distribution of *Ty1* retrotransposition in yeast (Freiberg et al. 2000; Ulrich 2002). Demonstrating an association between RecQ helicases and LTR retrotransposition, Sgs1 was found to be a repressor of *Ty1* retromobility (Bryk et al. 2001) and an increase of cancer predisposition was observed in BLM-deficient mice in response to challenge with MLV (Goss et al. 2002).

4.2.2
Non-LTR Retrotransposons

Non-LTR retrotransposons are mobilized by a process termed target-site primed reverse transcription (TPRT) (Fig. 10) that uses both an endonuclease (EN) (Feng et al. 1996) and an RT (Cost et al. 2002; Luan et al. 1993). During TPRT, one strand of host DNA is cleaved by the EN, which is then used by the RT as a primer in reverse transcription, copying the retrotransposon RNA template directly into the host genome. Non-LTR retrotransposons seem to be able to integrate only the 3′ end of their cDNA, while the 5′ integration needs an additional host factor (Fujimoto et al. 2004) and is thought to be facilitated by HDR. The TPRT model predicts that both strands of the genomic DNA are required to be cleaved upon integration. In agreement with this, Gasior et al. detected massive DSB signaling during *L1* retrotransposition (Gasior et al. 2006). Since microhomologies are frequently found at the 5′ joints, a microhomology-dependent repair mechanism might also be accounted for in certain cases. Strikingly, while in many organisms telomerase acts to elongate the ends of linear chromosomes, in *Drosophila* the telomere maintenance is carried out by a specialized RT of non-LTR retrotransposons, HeT-A and TART (Ahmad and Golic 1998; Biessmann et al. 1992a,b; Melnikova et al. 2005; Pardue et al. 1996). Furthermore, this process is regulated by the Ku70/80 factor of NHEJ (Melnikova et al. 2005). Interspersed nuclear elements, LINEs and SINEs, and long interspersed element-1 (*LINE* or *L1*) elements are abundant, non-LTR retrotransposons that comprise approximately 17% of human DNA. A subfamily of *L1* elements actively transposes in human populations (Sassaman et al. 1997) with a frequency of 1/50 individuals (Kazazian 2004). Short interspersed nuclear elements (SINEs) are nonautonomous, and rely on the activity of *LINEs* to retrotranspose (Dewannieux et al. 2003). Consequently, *L1*-encoded proteins mobilize other cellular RNAs to new genomic locations (Fig. 11b) (e.g., *Alu* retrotransposons, *SVA* retrotransposons, and U6 snRNAs), which comprise approximately 13% of human DNA (Garcia-Perez et al. 2007). This cross-mobilization has a mutagenic effect, as a new *Alu* insertion is estimated to occur once in every 20 individuals in humans (Cordaux et al. 2006). Similarly to LTR retrotranspo-

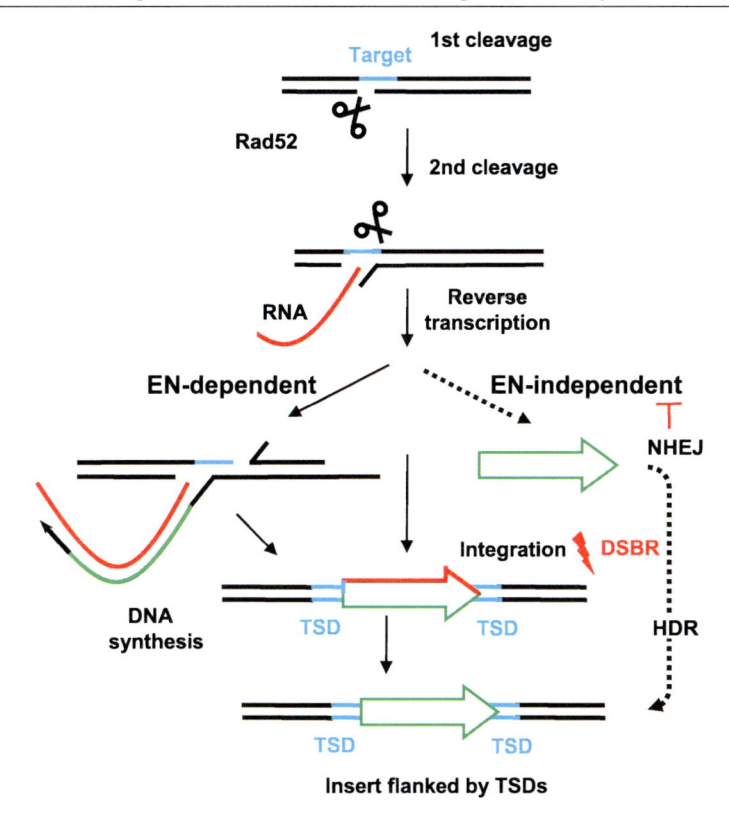

Fig. 10 Non-LTR retrotransposition. Non-LTR retrotransposons are mobilized by a process termed target-site primed reverse transcription (TPRT). During this process, one strand of host DNA is cleaved by the EN, which is then used by the RT as a primer in reverse transcription, copying the retrotransposon RNA template directly into the host genome. Non-LTR retrotransposons seem to be able to integrate only the 3′ end of their cDNA, while integration of the 5′ end is dependent on HDR. Upon integration, both strands of the genomic DNA are required to be cleaved, alerting DSB repair. The genomic integration of a retrotransposon can occur by EN-dependent or -independent pathways. The EN-independent process requires only the presence of cDNA, and is suppressed by NHEJ in mammalian cells by negatively influencing the fate of the free cDNA molecules. Similarly to LTR retroposons, *L*1 was also reported to integrate into preexisting *L*1 elements by the HDR process

sition, a "false" transposition pathway exists for non-LTR retrotransposition as well (Fig. 10). This EN-independent process requires only the presence of cDNA. This EN-independent pathway is suppressed by NHEJ in mammalian cells by negatively influencing the fate of the free cDNA molecules (Morrish et al. 2002). Similarly to LTR retrotransposons, *L*1 was also reported to integrate into preexisting *L*1 elements by the HDR process (Gilbert et al. 2002; Symer et al. 2002).

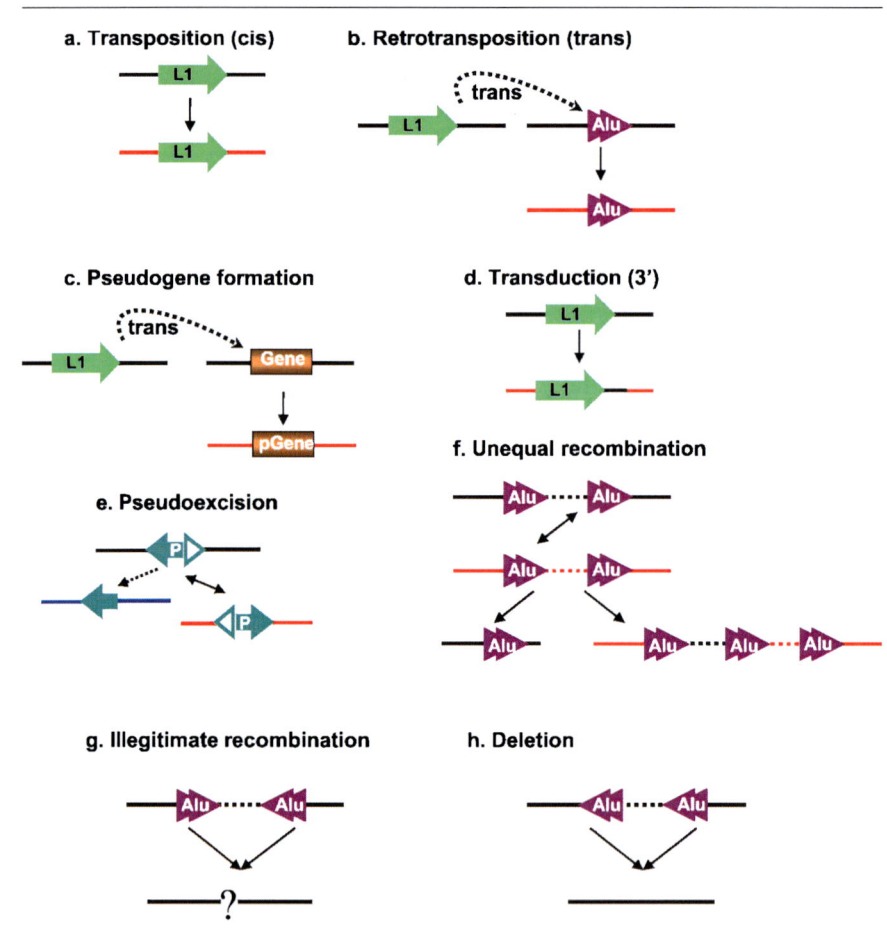

Fig. 11 Repetitive elements and genome stability. Selected examples of genome rearrangements associated by repetitive elements. **a** The mobilization of the element has a potential of insertional mutagenesis. **b** The element (*L1*) mobilizes other elements (*Alu*) in trans. **c** Trans-mobilization of other nonautonomous RNAs might contribute to pseudogene formation. **d** The *L1* element can transduce flanking genomic DNA to new locations when *L1*RNA contains upstream or downstream sequences. **e** "Pseudo-excision" during *P element* transposition: a combination of two deleted *P elements* in *Drosophila*, each containing one functional end, can be cleaved. During transposition without proper assembly of the paired-end complex (*dashed arrow*), the second end of the transposon is cleaved inside or outside (transduction—not shown) of the element. **f** Unequal recombination between existing elements in the genome can result in chromosomal duplications and deletions. Genome instability by the presence of **g** inverted or **h** directly repeated elements in the genome

Group II introns are mobilized by a process called retrohoming. During retrohoming the intron RNA reverse splices directly into one strand of a DNA target site and is then reverse transcribed by the associated intron-encoded protein. The completion of retrohoming is viewed as a DNA repair process, with features shared by other non-LTR retroelements (Smith et al. 2005).

5
Repetitive Elements and Genome Stability

Both DNA transposons and retroelements contribute significantly to chromosomal rearrangements. Imprecise transposition events have long been assumed to be responsible for restructuring the chromosomes (Gray 2000; Lim and Simmons 1994). Since aberrant transposition reactions are highly recombinogenic, transposition reactions are strictly regulated. Such an example is transposition of the *Mu* element (Mizuuchi et al. 1995; Pathania et al. 2002; Pato 1989) or V(D)J recombination (Jones and Gellert 2002), where the early phase of the reaction is tightly regulated. The ordered process of DNA recombinase protein assembly efficiently filters out aberrant transposition intermediates. In contrast, transposition of the *mariner* or *P elements* appears to be less stringently controlled. Nicking, and probably cleavage, can occur in *mariner* transposition at a single IR, without proper assembly of the paired-end complex (Dawson and Finnegan 2003; Lipkow et al. 2004). In such cases, the second cleavage that liberates the transposon can occur internally inside the transposon, or outside of the elements, with a potential of mobilizing genomic sequences. In agreement with this, several types of DNA transposons were reported to frequently transduce gene fragments in grass genomes (Lisch 2002). Another example of pseudo-excision was observed during *P element* transposition: a combination of two deleted *P elements* in *Drosophila*, each containing one functional end, was detected to be cleaved (Fig. 11e). It was demonstrated that the two ends, located even on different chromosomes, can occasionally associate and be cleaved as a hybrid transposon, causative of high levels of recombination (Gray et al. 1996). These pseudo-transposition events can lead to additional rearrangements, including chromosomal inversions, duplications, and deletions of several kilobases (Gray 2000; Lim and Simmons 1994). Transposase-induced but aberrant transposition can explain the occurrence of chromosome rearrangements in diverse organisms (Engels and Preston 1984; Hua-Van et al. 2002; Walker et al. 1995; Weil and Wessler 1993). In addition, these events can mediate exon shuffling and create new chimeric functional genes (Zhang et al. 2006).

Upon excision, DNA transposons can also generate mutations. Several transposons generate footprints at the excision site. Furthermore, excision of *hAT* elements and V(D)J recombination generates a hairpin at the donor site (Fig. 5b). The resolution of this hairpin generates a great deal of genetic variation (Gellert 2002) (Fig. 5b). This feature contributes to immunoglobulin diversity after the excision of the paralyzed element in V(D)J recombination. Similarly, the *hAT* family of transposons also leaves a hairpin structure at the excision site. Frequent excision of a *Tol2* element integrated in the promoter region of a pigmentation gene results in a wide range of pigmentation phenotypes in the medaka fish, ranging from albino to wild type through partially pigmented patterns (Koga et al. 2006). In general, DSBs generated by any transposase are exposed to the host repair machinery; the excision site often includes small deletions, inversions, or even short stretches of random filler DNA.

Several de novo integration events occur in the close proximity of the donor site. This phenomenon, called "local hopping", was observed in many transposition reactions and was suggested to generate complex, unstable, and highly recombigenic genomic structures (Geurts et al. 2006; Guimond et al. 2003; Keng et al. 2005; Moreno et al. 1992; Tower et al. 1993).

In addition to insertional mutagenesis, several retrotransposition events have "secondary" consequences as well. For example, *L1* and *Ty1* transposition were observed to be frequently followed by rearrangements (Gilbert et al. 2005; Pickeral et al. 2000; Symer et al. 2002) including the generation of chimeric pseudogenes (Fig. 10d) (Maxwell and Curcio 2007), intrachromosomal deletions (Gilbert et al. 2002), intrachromosomal duplications, and interchromosomal translocations (Fig. 11f). Both *Ty1* (Moore and Haber 1996) and *L1* transposition are facilitated by already existing gaps, generated by environmental stressors, such as oxidative stress (Rockwood et al. 2004), heavy metals, and gamma radiation (Farkash et al. 2006). Retroelements can reverse transcribe, and integrate cellular mRNAs and generate pseudogenes, having a significant effect in genome evolution (Carlton et al. 1995; Moran et al. 1999; Morrish et al. 2002).

Transposable elements are present by several copies in genomes. In addition to mobilization-related mutagenic effects, the simple presence of transposon-derived sequences can lead to a variety of rearrangements in the genome (Chen et al. 2005; Ostertag et al. 2003). SINEs, represented by extremely high copy numbers, are real challenges for the host repair. There are more than a million copies of *Alu* in the human genome, and indeed rearrangements frequently occur at or near *Alu* sequences (Fig. 11f,g) (Deininger and Batzer 1999; Kolomietz et al. 2002; Weinstock et al. 2006). The rearrangements could happen both in somatic and germline cells and can cause a variety of diseases. There are several examples of *Alu*-associated genome rearrangements, resulting in disease phenotypes (Rossetti et al. 2004; Onno et al. 1992; Schichman et al. 1994). In general, repetitive element rearrangements include deletions, sequence duplications, inversions, or translocations (Kolomietz et al. 2002; Sen et al. 2006).

It is believed that most translocations are initiated by formation of DSBs (Richardson and Jasin 2000) in the vicinity of repetitive elements. Highly similar, repeated sequences frequently initiate homologous recombination. SSA deletes sequences between highly similar, directly repeated motifs. Transposons are generally framed by short (2–20 bp), directly repeated TSDs generated upon transposition. For the benefit of the genome, SSA can function as a "genome purifier" and efficiently delete transposons between TSDs. However, in addition to genome purification, SSA can delete genomic sequences located between two nearly identical repetitive elements (van de Lagemaat et al. 2005). Importantly, sequence divergence significantly suppresses SSA or HDR in mammalian cells. Notably, host defense mechanisms aim to decrease the degree of homology between repetitive elements. Repeat-induced point mutation (RIP) is a homology-based process that mutates repetitive sequences (Galagan and Selker 2004).

Some fragile genomic regions are clearly associated with recognition motifs of endogenous recombination enzymes. The misrecognition of these motifs might lead to translocations or deletions. It was hypothesized that certain fragile sites, including LMO-2, represent cryptic sites (pseudo-signals) for V(D)J recombination that are occasionally recognized and treated as normal recombination signal sequences (RSSs) by the RAG1/2 recombinases (reviewed in Raghavan and Lieber 2006). In addition to targeting failure, other types of aberrant V(D)J recombination events, such as nicking and repair errors, were reported to generate chromosomal rearrangements (Lieber et al. 2006; Marculescu et al. 2006; Vanura et al. 2007). RAG-mediated transposition may also contribute significantly to genomic instability. It was estimated to occur approximately once every 50 000 V(D)J recombinations (Reddy et al. 2006).

Transposons are usually associated with unusual secondary structures (inverted and direct repeats). Recent studies demonstrated a role for unusual DNA structures, non-B DNA structures, as initiating factors to create DSBs at fragile genomic sites (Raghavan and Lieber 2006). These non-B DNA structures are vulnerable to structure-specific nucleases. Transposon-derived sequences, especially certain combinations, can form non-B DNA structures and be highly recombigenic. As a working model, head to head arrangements of Ty elements are preferred sites for DSBs, faithfully mimicking fragile sites in human chromosomes (Lemoine et al. 2005). The *Foldback* (FB) element of *Drosophila* possesses a large (often over 10 kb) inverted repeat structure, provoking an extremely high frequency of rearrangements (Collins and Rubin 1984).

6
Transposition and Cell Cycle

The successful propagation of transposable elements requires the host DNA repair machinery, which is employed in a cell cycle phase-specific way. NHEJ acts predominantly for DSB repair in the G1 to the early S phase of the cell cycle, while HDR is mainly employed in the late S to G2 phase (Fig. 12) (reviewed in Rodrigue et al. 2006).

Infection by oncoretroviruses, such as Murine Leukemia Virus (MLV) and Rous Sarcoma Virus (RSV), requires cell division (Roe et al. 1993). In nondividing cells, viral entry, uncoating, DNA synthesis, and formation of viral preintegration complexes (PICs) occur at the same rate as in dividing cells, but integration fails to occur (Roe et al. 1993). It has been thought that the physical size of the PIC exceeds the upper limit for passive diffusion through nuclear pores but, during mitosis, the nuclear membrane disassembles, rendering the chromosomes accessible to the PIC. However, imposed nuclear translocation of MLV DNA in nondividing cells is not sufficient for stable transduction (Lieber et al. 2000). It appears, therefore, that additional cellular factors activated during S phase or DNA repair synthesis are required for efficient retroviral integration. There are controversial reports whether or not L1 retrotransposition requires cell division (Kubo et al. 2006; Shi et al. 2007). In V(D)J recombination, selective degradation of the RAG2 protein has been proposed to contribute to confining V(D)J recombination to the G1 phase of the cell cycle (Li et al. 1996), in which the NHEJ pathway is preferentially active (Takata et al. 1998). RAG2 accumulates in G0/G1, declines before the

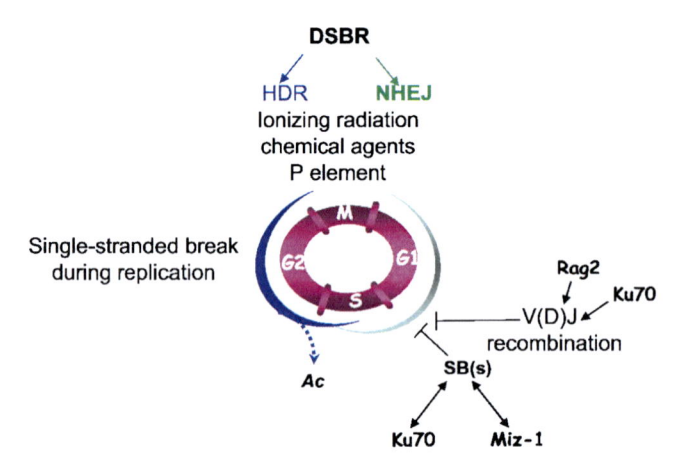

Fig. 12 Transposition and cell cycle. NHEJ acts predominantly for DSB repair in the G1 to the early S phases of the cell cycle, whereas HDR is mainly employed in the late S to G2 phases. In V(D)J recombination, selective degradation of the RAG2 protein contributes to restricting V(D)J recombination to the G1 phase of the cell cycle, in which the NHEJ pathway is preferentially active. RAG2 accumulates in G0/G1, declines before the cell enters the S phase, and remains low throughout the rest of the cell cycle. In somatic cells, *Sleeping Beauty* transposition is associated with NHEJ, promoted by two mechanisms. First, SB transposase physically interacts with Ku70 that is predominant in the G1 phase of the cell cycle. Second, the *Sleeping Beauty* transposase modulates the cell-cycle profile, resulting in a temporary arrest in the G1 phase via an interaction with Miz1. *P element* transposition does not seem to be associated to any particular cell-cycle phase. Transposition of the *Ac* element in maize from a replicated donor site to both unreplicated and replicated target sites implies that most transpositions of *Ac* occur during or shortly after the S phase of the cell cycle

cell enters the S phase, and remains low throughout the rest of the cell cycle (Fig. 12) (Lin and Desiderio 1994).

Transposition of the *Ac* element in maize from a replicated donor site to both unreplicated and replicated target sites implies that most transpositions of *Ac* occur during or shortly after the S phase of the cell cycle (Fig. 12) (Chen et al. 1992). In contrast, *P element* transposition does not seem to be associated with any particular cell-cycle phase (Weinert et al. 2005). It is currently unknown whether other DNA-based transposable elements, such as *Sleeping Beauty*, require active cell division for transposition and, if yes, whether transposition can occur throughout the cell cycle or is restricted to certain cell-cycle phases. Two lines of observation indicate that in somatic cells, *Sleeping Beauty* transposition is associated with the G1 phase of the cell cycle. First, SB transposase physically interacts with Ku70 (Izsvák et al. 2004), a key component of the NHEJ repair pathway that is predominant in the G1 phase of the cell cycle. The *Sleeping Beauty* transposon was also shown to modulate the cell-cycle profile, resulting in a temporary arrest in the G1 phase via an interaction with Miz1 (Walisko et al. 2006). However, in contrast to V(D)J recombination, *Tc1* transposons amplify in copy number, which implies that under certain conditions transposition should occur in cell-cycle phases other than G1. Thus, *Sleeping Beauty* should have a different strategy to collaborate with HDR pathways, which are preferentially active in the S and G2 phases of the cell cycle.

7
Concluding Remarks

In order to preserve genome integrity, the cell is a permanent battlefield between transposons and host defense. From the host's point of view, the interaction with the transposons can be "offensive" or "defensive". In the defensive hypothesis, described for *Ty1*, transposable elements (Sawyer and Malik 2006) use the repair factors to localize already existing DSBs to integrate. Thus, in order to maximize host fitness, repair proteins would evolve "away" from *Ty1* recognition. In the alternative offensive model, transposition intermediates are neutralized by repair proteins, which circularize linear *Ty1* molecules into dead-end products. In this model, the repair proteins are evolving "toward" recognition of *Ty1* transposition intermediates.

As we have seen, the interaction between transposons and the host repair machinery is complex. The host has developed inhibitory strategies against transposition and repair signaling mechanisms ("genome caretakers") that are also part of the host defense. In the absence of these mechanisms, transposon mobilization is more frequent. There are cellular mechanisms that are piggybacked by transposons, including DNA repair, to support transposition or even transposon amplification. Transposition is thought to have an intimate relationship with stress and have different strategies to sense changes in the environment. In response to environmental signals, cellular processes affecting repair and differentiation seem to regulate levels of transposition (Curcio and Garfinkel 1999; Scholes et al. 2003). As Barbara McClintock first hypothesized, transposition may contribute to genotoxic stress and conversely, genotoxic stress may contribute to the mobilization of transposable elements (McClintock 1984a).

Acknowledgements Work in the authors' laboratory was supported by a grant from the Deutsche Forschungsgemeinschaft IV 21/3-1.

References

Adams MD, McVey M, Sekelsky JJ (2003) Drosophila BLM in double-strand break repair by synthesis-dependent strand annealing. Science 299:265–267

Ahmad K, Golic KG (1998) The transmission of fragmented chromosomes in *Drosophila melanogaster*. Genetics 148:775–792

Alleman M, Freeling M (1986) The Mu transposable elements of maize: evidence for transposition and copy number regulation during development. Genetics 112:107–119

Arca B, Zabalou S, Loukeris TG, Savakis C (1997) Mobilization of a Minos transposon in *Drosophila melanogaster* chromosomes and chromatid repair by heteroduplex formation. Genetics 145:267–279

Arcangioli B (2000) Fate of mat1 DNA strands during mating-type switching in fission yeast. EMBO Rep 1:145–150

Baumann P, West SC (1998) Role of the human RAD51 protein in homologous recombination and double-stranded-break repair. Trends Biochem Sci 23:247–251

Beall EL, Rio DC (1996) Drosophila IRBP/Ku p70 corresponds to the mutagen-sensitive mus309 gene and is involved in P-element excision in vivo. Genes Dev 10:921–933

Beall EL, Mahoney MB, Rio DC (2002) Identification and analysis of a hyperactive mutant form of Drosophila P-element transposase. Genetics 162:217–227

Biessmann H, Champion LE, O'Hair M, Ikenaga K, Kasravi B, Mason JM (1992a) Frequent transpositions of *Drosophila melanogaster* HeT-A transposable elements to receding chromosome ends. EMBO J 11:4459–4469

Biessmann H et al (1992b) HeT-A, a transposable element specifically involved in "healing" broken chromosome ends in *Drosophila melanogaster*. Mol Cell Biol 12:3910–3918

Broomfield S, Hryciw T, Xiao W (2001) DNA postreplication repair and mutagenesis in *Saccharomyces cerevisiae*. Mutat Res 486:167–184

Bryk M, Banerjee M, Conte D Jr, Curcio MJ (2001) The Sgs1 helicase of *Saccharomyces cerevisiae* inhibits retrotransposition of Ty1 multimeric arrays. Mol Cell Biol 21:5374–5388

Burma S, Chen BP, Chen DJ (2006) Role of nonhomologous end joining (NHEJ) in maintaining genomic integrity. DNA Repair (Amst) 5:1042–1048

Calvi BR, Hong TJ, Findley SD, Gelbart WM (1991) Evidence for a common evolutionary origin of inverted repeat transposons in Drosophila and plants: hobo, Activator, and Tam3. Cell 66:465–471

Carlton MB, Colledge WH, Evans MJ (1995) Generation of a pseudogene during retroviral infection. Mamm Genome 6:90–95

Chen J, Greenblatt IM, Dellaporta SL (1987) Transposition of Ac from the P locus of maize into unreplicated chromosomal sites. Genetics 117:109–116

Chen J, Greenblatt IM, Dellaporta SL (1992) Molecular analysis of Ac transposition and DNA replication. Genetics 130:665–676

Chen JM, Stenson PD, Cooper DN, Ferec C (2005) A systematic analysis of LINE-1 endonuclease-dependent retrotranspositional events causing human genetic disease. Hum Genet 117:411–427

Collins M, Rubin GM (1984) Structure of chromosomal rearrangements induced by the FB transposable element in Drosophila. Nature 308:323–327

Colot V, Haedens V, Rossignol JL (1998) Extensive, nonrandom diversity of excision footprints generated by Ds-like transposon Ascot-1 suggests new parallels with V(D)J recombination. Mol Cell Biol 18:4337–4346

Cordaux R, Hedges DJ, Herke SW, Batzer MA (2006) Estimating the retrotransposition rate of human Alu elements. Gene 373:134–137

Corneo B et al (2007) Rag mutations reveal robust alternative end joining. Nature 449:483–486

Cost GJ, Feng Q, Jacquier A, Boeke JD (2002) Human L1 element target-primed reverse transcription in vitro. EMBO J 21:5899–5910

Couedel C et al (2004) Collaboration of homologous recombination and nonhomologous end-joining factors for the survival and integrity of mice and cells. Genes Dev 18:1293–1304

Craig NL (1995) Unity in transposition reactions. Science 270:253–254

Curcio MJ, Garfinkel DJ (1999) New lines of host defense: inhibition of Ty1 retrotransposition by Fus3p and NER/TFIIH. Trends Genet 15:43–45

Curcio MJ et al (2007) S-phase checkpoint pathways stimulate the mobility of the retrovirus-like transposon Ty1. Mol Cell Biol 27:8874–8885

Daley JM, Palmbos PL, Wu D, Wilson TE (2005) Nonhomologous end joining in yeast. Annu Rev Genet 39:431–451

Daniel R et al (2003) Evidence that the retroviral DNA integration process triggers an ATR-dependent DNA damage response. Proc Natl Acad Sci USA 100:4778–4783

Dawson A, Finnegan DJ (2003) Excision of the Drosophila mariner transposon Mos1. Comparison with bacterial transposition and V(D)J recombination. Mol Cell 11:225–235

Deininger PL, Batzer MA (1999) Alu repeats and human disease. Mol Genet Metab 67:183–193

Delattre M, Anxolabehere D, Coen D (1995) Prevalence of localized rearrangements vs. transpositions among events induced by Drosophila P element transposase on a P transgene. Genetics 141:1407–1424

Dewannieux M, Esnault C, Heidmann T (2003) LINE-mediated retrotransposition of marked Alu sequences. Nat Genet 35:41–48

Dooner HK, Martinez-Ferez IM (1997) Germinal excisions of the maize transposon activator do not stimulate meiotic recombination or homology-dependent repair at the bz locus. Genetics 147:1923–1932

Doring HP, Starlinger P (1986) Molecular genetics of transposable elements in plants. Annu Rev Genet 20:175–200

Doring HP, Tillmann E, Starlinger P (1984) DNA sequence of the maize transposable element Dissociation. Nature 307:127–130

Downs JA, Jackson SP (1999) Involvement of DNA end-binding protein Ku in Ty element retrotransposition. Mol Cell Biol 19:6260–6268

Durocher D, Jackson SP (2001) DNA-PK, ATM and ATR as sensors of DNA damage: variations on a theme? Curr Opin Cell Biol 13:225–231

Egel R, Kohli J, Thuriaux P, Wolf K (1980) Genetics of the fission yeast *Schizosaccharomyces pombe*. Annu Rev Genet 14:77–108

Ehmsen KT, Heyer WD (2008) Biochemistry of Meiotic Recombination: Formation, Processing, and Resolution of Recombination Intermediates. In: Egel R, Lankenau DH (eds) Recombination and Meiosis – Models, Means and Evolution, vol 3. Springer, Berlin, Heidelberg, pp 91–164

Engels WR, Preston CR (1984) Formation of chromosome rearrangements by P factors in Drosophila. Genetics 107:657–678

Engels WR, Johnson-Schlitz DM, Eggleston WB, Sved J (1990) High-frequency P element loss in Drosophila is homolog dependent. Cell 62:515–525

Engels WR, Preston CR, Johnson-Schlitz DM (1994) Long-range cis preference in DNA homology search over the length of a Drosophila chromosome. Science 263:1623–1625

Farkash EA, Kao GD, Horman SR, Prak ET (2006) Gamma radiation increases endonuclease-dependent L1 retrotransposition in a cultured cell assay. Nucleic Acids Res 34:1196–1204

Feldmar S, Kunze R (1991) The ORFa protein, the putative transposase of maize transposable element Ac, has a basic DNA binding domain. EMBO J 10:4003–4010

Feng Q, Moran JV, Kazazian HH Jr, Boeke JD (1996) Human L1 retrotransposon encodes a conserved endonuclease required for retrotransposition. Cell 87:905–916

Feschotte C, Mouches C (2000) Evidence that a family of miniature inverted-repeat transposable elements (MITEs) from the *Arabidopsis thaliana* genome has arisen from a pogo-like DNA transposon. Mol Biol Evol 17:730–737

Feschotte C, Pritham EJ (2007) DNA transposons and the evolution of eukaryotic genomes. Annu Rev Genet 41:331–368

Fischer SE, Wienholds E, Plasterk RH (2003) Continuous exchange of sequence information between dispersed Tc1 transposons in the *Caenorhabditis elegans* genome. Genetics 164:127–134

Formosa T, Alberts BM (1986) DNA synthesis dependent on genetic recombination: characterization of a reaction catalyzed by purified bacteriophage T4 proteins. Cell 47:793–806

Franco S, Alt FW, Manis JP (2006) Pathways that suppress programmed DNA breaks from progressing to chromosomal breaks and translocations. DNA Repair (Amst) 5:1030–1041

Freiberg G, Mesecar AD, Huang H, Hong JY, Liebman SW (2000) Characterization of novel rad6/ubc2 ubiquitin-conjugating enzyme mutants in yeast. Curr Genet 37:221–233

Fujimoto H et al (2004) Integration of the 5′ end of the retrotransposon, R2Bm, can be complemented by homologous recombination. Nucleic Acids Res 32:1555–1565

Galagan JE, Selker EU (2004) RIP: the evolutionary cost of genome defense. Trends Genet 20:417–423

Garcia-Perez JL, Doucet AJ, Bucheton A, Moran JV, Gilbert N (2007) Distinct mechanisms for trans-mediated mobilization of cellular RNAs by the LINE-1 reverse transcriptase. Genome Res 17:602–611

Gasior SL, Wakeman TP, Xu B, Deininger PL (2006) The human LINE-1 retrotransposon creates DNA double-strand breaks. J Mol Biol 357:1383–1393

Gellert M (2002) V(D)J recombination: RAG proteins, repair factors, and regulation. Annu Rev Biochem 71:101–132

Geurts AM et al (2006) Gene mutations and genomic rearrangements in the mouse as a result of transposon mobilization from chromosomal concatemers. PLoS Genet 2:e156

Gilbert N, Lutz-Prigge S, Moran JV (2002) Genomic deletions created upon LINE-1 retrotransposition. Cell 110:315–325

Gilbert N, Lutz S, Morrish TA, Moran JV (2005) Multiple fates of L1 retrotransposition intermediates in cultured human cells. Mol Cell Biol 25:7780–7795

Gloor GB, Lankenau DH (1998) Gene conversion in mitotically dividing cells: a view from Drosophila. Trends Genet 14:43–46

Gloor GB, Nassif NA, Johnson-Schlitz DM, Preston CR, Engels WR (1991) Targeted gene replacement in Drosophila via P element-induced gap repair. Science 253:1110–1117

Gloor GB, Moretti J, Mouyal J, Keeler KJ (2000) Distinct P-element excision products in somatic and germline cells of *Drosophila melanogaster*. Genetics 155:1821–1830

Gorbunova V, Levy AA (1997) Circularized Ac/Ds transposons: formation, structure and fate. Genetics 145:1161–1169

Goss KH et al (2002) Enhanced tumor formation in mice heterozygous for Blm mutation. Science 297:2051–2053

Gray YH (2000) It takes two transposons to tango: transposable-element-mediated chromosomal rearrangements. Trends Genet 16:461–468

Gray YH, Tanaka MM, Sved JA (1996) P-element-induced recombination in *Drosophila melanogaster*: hybrid element insertion. Genetics 144:1601–1610

Guimond N, Bideshi DK, Pinkerton AC, Atkinson PW, O'Brochta DA (2003) Patterns of Hermes transposition in *Drosophila melanogaster*. Mol Genet Genomics 268:779–790

Haber JE (1992) Exploring the pathways of homologous recombination. Curr Opin Cell Biol 4:401–412

Haber JE (2000) Partners and pathways repairing a double-strand break. Trends Genet 16:259–264

Harper JW, Elledge SJ (2007) The DNA damage response: ten years after. Mol Cell 28:739–745

Hegde ML, Hazra TK, Mitra S (2008) Early steps in the DNA base excision/single-strand interruption repair pathway in mammalian cells. Cell Res 18:27–47

Heslip TR, Williams JA, Bell JB, Hodgetts RB (1992) A P element chimera containing captured genomic sequences was recovered at the vestigial locus in Drosophila following targeted transposition. Genetics 131:917–927

Hicks JB, Strathern J, Herskowitz I (1977) The cassette model of mating-type interconversion. In: Bukhari AI, Sahpiro SA, Adhya S (eds) DNA insertion elements. Cold Spring Harbor Laboratory, New York, pp 457–462

Hickson ID (2003) RecQ helicases: caretakers of the genome. Nat Rev Cancer 3:169–178

Hoege C, Pfander B, Moldovan GL, Pyrowolakis G, Jentsch S (2002) RAD6-dependent DNA repair is linked to modification of PCNA by ubiquitin and SUMO. Nature 419:135–141

Hsia AP, Schnable PS (1996) DNA sequence analyses support the role of interrupted gap repair in the origin of internal deletions of the maize transposon, MuDR. Genetics 142:603–618

Hua-Van A, Langin T, Daboussi MJ (2002) Aberrant transposition of a Tc1-mariner element, impala, in the fungus *Fusarium oxysporum*. Mol Genet Genomics 267:79–87

Hubscher U, Maga G, Spadari S (2002) Eukaryotic DNA polymerases. Annu Rev Biochem 71:133–163

International Human Genome Sequencing Consortium (2001) Initial sequencing and analysis of the human genome. Nature 409:860–921

Ivics Z, Hackett PB, Plasterk RH, Izsvák Z (1997) Molecular reconstruction of Sleeping Beauty, a Tc1-like transposon from fish, and its transposition in human cells. Cell 91:501–510

Izsvák Z, Ivics Z, Hackett PB (1995) Characterization of a Tc1-like transposable element in zebrafish (*Danio rerio*). Mol Gen Genet 247:312–322

Izsvák Z, Ivics Z, Plasterk RH (2000) Sleeping Beauty, a wide host-range transposon vector for genetic transformation in vertebrates. J Mol Biol 302:93–102

Izsvák Z, Stuwe EE, Fiedler D, Katzer A, Jeggo PA, Ivics Z (2004) Healing the wounds inflicted by Sleeping Beauty transposition by double-strand break repair in mammalian somatic cells. Mol Cell 13:279–290

Jackson SP, Jeggo PA (1995) DNA double-strand break repair and V(D)J recombination: involvement of DNA-PK. Trends Biochem Sci 20:412–415

Jasinskiene N et al (1998) Stable transformation of the yellow fever mosquito, *Aedes aegypti*, with the Hermes element from the housefly. Proc Natl Acad Sci USA 95:3743–3747

Jeggo P, O'Neill P (2002) The Greek goddess, Artemis, reveals the secrets of her cleavage. DNA Repair (Amst) 1:771–777

Johnson RD, Jasin M (2000) Sister chromatid gene conversion is a prominent double-strand break repair pathway in mammalian cells. EMBO J 19:3398–3407

Johnson-Schlitz DM, Engels WR (1993) P-element-induced interallelic gene conversion of insertions and deletions in *Drosophila melanogaster*. Mol Cell Biol 13:7006–7018

Jones JM, Gellert M (2002) Ordered assembly of the V(D)J synaptic complex ensures accurate recombination. EMBO J 21:4162–4171

Kapitonov VV, Jurka J (2005) RAG1 core and V(D)J recombination signal sequences were derived from Transib transposons. PLoS Biol 3:e181

Kawakami K, Koga A, Hori H, Shima A (1998) Excision of the tol2 transposable element of the medaka fish, *Oryzias latipes*, in zebrafish, *Danio rerio*. Gene 225:17–22

Kazazian HH Jr (2004) Mobile elements: drivers of genome evolution. Science 303:1626–1632

Keeney S (2008) Spo11 and the Formation of DNA Double-Strand Breaks in Meiosis. In: Egel R, Lankenau DH (eds) Recombination and Meiosis: Crossing-Over and Disjunction, vol 2. Springer, Heidelberg, pp 81–123

Keeney S, Neale MJ (2006) Initiation of meiotic recombination by formation of DNA double-strand breaks: mechanism and regulation. Biochem Soc Trans 34:523–525

Keng VW et al (2005) Region-specific saturation germline mutagenesis in mice using the Sleeping Beauty transposon system. Nat Methods 2:763–769

Kilzer JM, Stracker T, Beitzel B, Meek K, Weitzman M, Bushman FD (2003) Roles of host cell factors in circularization of retroviral DNA. Virology 314:460–467

Koga A, Iida A, Hori H, Shimada A, Shima A (2006) Vertebrate DNA transposon as a natural mutator: the medaka fish Tol2 element contributes to genetic variation without recognizable traces. Mol Biol Evol 23:1414–1419

Kolomietz E, Meyn MS, Pandita A, Squire JA (2002) The role of Alu repeat clusters as mediators of recurrent chromosomal aberrations in tumors. Genes Chromosomes Cancer 35:97–112

Kooistra R, Pastink A, Zonneveld JB, Lohman PH, Eeken JC (1999) The *Drosophila melanogaster* DmRAD54 gene plays a crucial role in double-strand break repair after P-element excision and acts synergistically with Ku70 in the repair of X-ray damage. Mol Cell Biol 19:6269–6275

Krogh BO, Symington LS (2004) Recombination proteins in yeast. Annu Rev Genet 38:233–271

Kubo S et al (2006) L1 retrotransposition in nondividing and primary human somatic cells. Proc Natl Acad Sci USA 103:8036–8041

Lankenau DH (1995) Genetics of genetics in Drosophila: P elements serving the study of homologous recombination, gene conversion and targeting. Chromosoma 103:659–668

Lankenau DH, Gloor GB (1998) In vivo gap repair in Drosophila: a one-way street with many destinations. Bioessays 20:317–327

Lankenau DH, Peluso MV, Lankenau S (2000) The Su(Hw) chromatin insulator protein alters double-strand break repair frequencies in the Drosophila germ line. Chromosoma 109:148–160

Lau A, Kanaar R, Jackson SP, O'Connor MJ (2004) Suppression of retroviral infection by the RAD52 DNA repair protein. EMBO J 23:3421–3429

Lau A et al. (2005) Suppression of HIV-1 infection by a small molecule inhibitor of the ATM kinase. Nat Cell Biol 7:493–500

Lavoie BD, Chaconas G (1996) Transposition of phage Mu DNA. Curr Top Microbiol Immunol 204:83–102

Lee BS, Bi L, Garfinkel DJ, Bailis AM (2000) Nucleotide excision repair/TFIIH helicases RAD3 and SSL2 inhibit short-sequence recombination and Ty1 retrotransposition by similar mechanisms. Mol Cell Biol 20:2436–2445

Lee GS, Neiditch MB, Salus SS, Roth DB (2004) RAG proteins shepherd double-strand breaks to a specific pathway, suppressing error-prone repair, but RAG nicking initiates homologous recombination. Cell 117:171–184

Lees-Miller SP, Meek K (2003) Repair of DNA double strand breaks by nonhomologous end joining. Biochimie 85:1161–1173

Lemoine FJ, Degtyareva NP, Lobachev K, Petes TD (2005) Chromosomal translocations in yeast induced by low levels of DNA polymerase: a model for chromosome fragile sites. Cell 120:587–598

Li GM (2008) Mechanisms and functions of DNA mismatch repair. Cell Res 18:85–98

Li J, Wen TJ, Schnable PS (2008) Role of RAD51 in the Repair of MuDR-Induced Double-Strand Breaks in Maize (*Zea mays* L.). Genetics 178:57–66

Li L et al (2001) Role of the nonhomologous DNA end joining pathway in the early steps of retroviral infection. EMBO J 20:3272–3281

Li Z, Dordai DI, Lee J, Desiderio S (1996) A conserved degradation signal regulates RAG-2 accumulation during cell division and links V(D)J recombination to the cell cycle. Immunity 5:575–589

Lieber A, Kay MA, Li ZY (2000) Nuclear import of moloney murine leukemia virus DNA mediated by adenovirus preterminal protein is not sufficient for efficient retroviral transduction in nondividing cells. J Virol 74:721–734

Lieber MR, Yu K, Raghavan SC (2006) Roles of nonhomologous DNA end joining, V(D)J recombination, and class switch recombination in chromosomal translocations. DNA Repair (Amst) 5:1234–1245

Lim JK, Simmons MJ (1994) Gross chromosome rearrangements mediated by transposable elements in *Drosophila melanogaster*. Bioessays 16:269–275

Lin CT, Lyu YL, Xiao H, Lin WH, Whang-Peng J (2001) Suppression of gene amplification and chromosomal DNA integration by the DNA mismatch repair system. Nucleic Acids Res 29:3304–3310

Lin WC, Desiderio S (1994) Cell cycle regulation of V(D)J recombination-activating protein RAG-2. Proc Natl Acad Sci USA 91:2733–2737

Lin Y, Waldman AS (2001) Capture of DNA sequences at double-strand breaks in mammalian chromosomes. Genetics 158:1665–1674

Lipkow K, Buisine N, Lampe DJ, Chalmers R (2004) Early intermediates of mariner transposition: catalysis without synapsis of the transposon ends suggests a novel architecture of the synaptic complex. Mol Cell Biol 24:8301–8311

Lisch D (2002) Mutator transposons. Trends Plant Sci 7:498–504

Luan DD, Korman MH, Jakubczak JL, Eickbush TH (1993) Reverse transcription of R2Bm RNA is primed by a nick at the chromosomal target site: a mechanism for non-LTR retrotransposition. Cell 72:595–605

Mahillon J, Chandler M (1998) Insertion sequences. Microbiol Mol Biol Rev 62:725–774

Marculescu R et al (2006) Recombinase, chromosomal translocations and lymphoid neoplasia: targeting mistakes and repair failures. DNA Repair (Amst) 5:1246–1258

Maxwell PH, Curcio MJ (2007) Retrosequence formation restructures the yeast genome. Genes Dev 21:3308–3318

May EW, Craig NL (1996) Switching from cut-and-paste to replicative Tn7 transposition. Science 272:401–404

Mazin AV, Alexeev AA, Kowalczykowski SC (2003) A novel function of Rad54 protein. Stabilization of the Rad51 nucleoprotein filament. J Biol Chem 278:14029–14036

McClintock B (1950) The origin and behavior of mutable loci in maize. Proc Natl Acad Sci USA 36:344–355

McClintock B (1984) The significance of responses of the genome to challenge. Science 226:792–801

McVey M, Adams M, Staeva-Vieira E, Sekelsky JJ (2004a) Evidence for multiple cycles of strand invasion during repair of double-strand gaps in Drosophila. Genetics 167:699–705

McVey M, Radut D, Sekelsky JJ (2004b) End-joining repair of double-strand breaks in Drosophila melanogaster is largely DNA ligase IV independent. Genetics 168:2067–2076

Melnikova L, Biessmann H, Georgiev P (2005) The Ku protein complex is involved in length regulation of Drosophila telomeres. Genetics 170:221–235

Min B, Weinert BT, Rio DC (2004) Interplay between Drosophila Bloom's syndrome helicase and Ku autoantigen during nonhomologous end joining repair of P element-induced DNA breaks. Proc Natl Acad Sci USA 101:8906–8911

Mitra R, Fain-Thornton J, Craig NL (2008) piggyBac can bypass DNA synthesis during cut and paste transposition. EMBO J 27:1097–1109

Mizuuchi M, Baker TA, Mizuuchi K (1995) Assembly of phage Mu transpososomes: cooperative transitions assisted by protein and DNA scaffolds. Cell 83:375–385

Moerman DG, Kiff JE, Waterston RH (1991) Germline excision of the transposable element Tc1 in C. elegans. Nucleic Acids Res 19:5669–5672

Moore JK, Haber JE (1996) Capture of retrotransposon DNA at the sites of chromosomal double-strand breaks. Nature 383:644–646

Moran JV, DeBerardinis RJ, Kazazian HH Jr (1999) Exon shuffling by L1 retrotransposition. Science 283:1530–1534

Moreno MA, Chen J, Greenblatt I, Dellaporta SL (1992) Reconstitutional mutagenesis of the maize P gene by short-range Ac transpositions. Genetics 131:939–956

Morrish TA et al (2002) DNA repair mediated by endonuclease-independent LINE-1 retrotransposition. Nat Genet 31:159–165

Moynahan ME, Jasin M (1997) Loss of heterozygosity induced by a chromosomal double-strand break. Proc Natl Acad Sci USA 94:8988–8993

Mulder LC, Chakrabarti LA, Muesing MA (2002) Interaction of HIV-1 integrase with DNA repair protein hRad18. J Biol Chem 277:27489–27493

Nassif N, Penney J, Pal S, Engels WR, Gloor GB (1994) Efficient copying of nonhomologous sequences from ectopic sites via P-element-induced gap repair. Mol Cell Biol 14:1613–1625

Nunnari G, Argyris E, Fang J, Mehlman KE, Pomerantz RJ, Daniel R (2005) Inhibition of HIV-1 replication by caffeine and caffeine-related methylxanthines. Virology 335:177–184

O'Brochta DA, Gomez SP, Handler AM (1991) P element excision in Drosophila melanogaster and related drosophilids. Mol Gen Genet 225:387–394

Onno M, Nakamura T, Hillova J, Hill M (1992) Rearrangement of the human tre oncogene by homologous recombination between Alu repeats of nucleotide sequences from two different chromosomes. Oncogene 7:2519–2523

Ostertag EM, Goodier JL, Zhang Y, Kazazian HH Jr (2003) SVA elements are nonautonomous retrotransposons that cause disease in humans. Am J Hum Genet 73:1444–1451

Paques F, Haber JE (1999) Multiple pathways of recombination induced by double-strand breaks in *Saccharomyces cerevisiae*. Microbiol Mol Biol Rev 63:349–404

Paques F, Leung WY, Haber JE (1998) Expansions and contractions in a tandem repeat induced by double-strand break repair. Mol Cell Biol 18:2045–2054

Pardue ML, Danilevskaya ON, Lowenhaupt K, Slot F, Traverse KL (1996) Drosophila telomeres: new views on chromosome evolution. Trends Genet 12:48–52

Pathania S, Jayaram M, Harshey RM (2002) Path of DNA within the Mu transpososome. Transposase interactions bridging two Mu ends and the enhancer trap five DNA supercoils. Cell 109:425–436

Pato ML (1989) Bacteriophage Mu23-52. In: Berg DE, Howe MM (eds) Mobile DNA, vol 1. American Society of Microbiology, Washington

Paunesku T et al (2001) Proliferating cell nuclear antigen (PCNA): ringmaster of the genome. Int J Radiat Biol 77:1007–1021

Pickeral OK, Makalowski W, Boguski MS, Boeke JD (2000) Frequent human genomic DNA transduction driven by LINE-1 retrotransposition. Genome Res 10:411–415

Plasterk RH, Groenen JT (1992) Targeted alterations of the *Caenorhabditis elegans* genome by transgene instructed DNA double strand break repair following Tc1 excision. EMBO J 11:287–290

Plasterk RH, Izsvák Z, Ivics Z (1999) Resident aliens: the Tc1/mariner superfamily of transposable elements. Trends Genet 15:326–332

Posey JE, Pytlos MJ, Sinden RR, Roth DB (2006) Target DNA structure plays a critical role in RAG transposition. PLoS Biol 4:e350

Preston CR, Flores C, Engels WR (2006a) Age-dependent usage of double-strand-break repair pathways. Curr Biol 16:2009–2015

Preston CR, Flores CC, Engels WR (2006b) Differential usage of alternative pathways of double-strand break repair in Drosophila. Genetics 172:1055–1068

Pritham EJ, Putliwala T, Feschotte C (2007) Mavericks, a novel class of giant transposable elements widespread in eukaryotes and related to DNA viruses. Gene 390:3–17

Puchta H (2005) The repair of double-strand breaks in plants: mechanisms and consequences for genome evolution. J Exp Bot 56:1–14

Radford SJ et al (2004) Increase in Ty1 cDNA recombination in yeast sir4 mutant strains at high temperature. Genetics 168:89–101

Radice AD, Emmons SW (1993) Extrachromosomal circular copies of the transposon Tc1. Nucleic Acids Res 21:2663–2667

Raghavan SC, Lieber MR (2006) DNA structures at chromosomal translocation sites. Bioessays 28:480–494

Rattray AJ, Shafer BK, Garfinkel DJ (2000) The *Saccharomyces cerevisiae* DNA recombination and repair functions of the RAD52 epistasis group inhibit Ty1 transposition. Genetics 154:543–556

Raval P, Kriatchko AN, Kumar S, Swanson PC (2008) Evidence for Ku70/Ku80 association with full-length RAG1. Nucleic Acids Res 36:2060–2072

Raynard S, Bussen W, Sung P (2006) A double Holliday junction dissolvasome comprising BLM, topoisomerase IIIalpha, and BLAP75. J Biol Chem 281:13861–13864

Reddy YV, Perkins EJ, Ramsden DA (2006) Genomic instability due to V(D)J recombination-associated transposition. Genes Dev 20:1575–1582

Resnick MA (1976) The repair of double-strand breaks in DNA; a model involving recombination. J Theor Biol 59:97–106

Richardson C, Jasin M (2000) Frequent chromosomal translocations induced by DNA double-strand breaks. Nature 405:697–700

Riley T, Sontag E, Chen P, Levine A (2008) Transcriptional control of human p53-regulated genes. Nat Rev Mol Cell Biol 9:402–412

Robert V, Bessereau JL (2007) Targeted engineering of the *Caenorhabditis elegans* genome following Mos1-triggered chromosomal breaks. EMBO J 26:170–183

Rockwood LD, Felix K, Janz S (2004) Elevated presence of retrotransposons at sites of DNA double strand break repair in mouse models of metabolic oxidative stress and MYC-induced lymphoma. Mutat Res 548:117–125

Rodrigue A et al (2006) Interplay between human DNA repair proteins at a unique double-strand break in vivo. EMBO J 25:222–231

Roe TY, Reynolds TC, Yu G, Brown PO (1993) Integration of murine leukemia virus DNA depends on mitosis. EMBO J 12:2099–2108

Rossetti LC, Goodeve A, Larripa IB, De Brasi CD (2004) Homologous recombination between AluSx-sequences as a cause of hemophilia. Hum Mutat 24:440

Roth DB (2003) Restraining the V(D)J recombinase. Nat Rev Immunol 3:656–666

Rothstein R, Michel B, Gangloff S (2000) Replication fork pausing and recombination or "gimme a break". Genes Dev 14:1–10

Rouse J, Jackson SP (2002) Interfaces between the detection, signaling, and repair of DNA damage. Science 297:547–551

Rubin E, Levy AA (1997) Abortive gap repair: underlying mechanism for Ds element formation. Mol Cell Biol 17:6294–6302

Sassaman DM et al (1997) Many human L1 elements are capable of retrotransposition. Nat Genet 16:37–43

Sawyer SL, Malik HS (2006) Positive selection of yeast nonhomologous end-joining genes and a retrotransposon conflict hypothesis. Proc Natl Acad Sci USA 103:17614–17619

Schichman SA et al (1994) ALL-1 tandem duplication in acute myeloid leukemia with a normal karyotype involves homologous recombination between Alu elements. Cancer Res 54:4277–4280

Scholes DT, Kenny AE, Gamache ER, Mou Z, Curcio MJ (2003) Activation of a LTR-retrotransposon by telomere erosion. Proc Natl Acad Sci USA 100:15736–15741

Scott L, LaFoe D, Weil CF (1996) Adjacent sequences influence DNA repair accompanying transposon excision in maize. Genetics 142:237–246

Sen SK et al (2006) Human genomic deletions mediated by recombination between Alu elements. Am J Hum Genet 79:41–53

Shapiro JA (1979) Molecular model for the transposition and replication of bacteriophage Mu and other transposable elements. Proc Natl Acad Sci USA 76:1933–1937

Sharon G, Burkett TJ, Garfinkel DJ (1994) Efficient homologous recombination of Ty1 element cDNA when integration is blocked. Mol Cell Biol 14:6540–6551

Shi X, Seluanov A, Gorbunova V (2007) Cell divisions are required for L1 retrotransposition. Mol Cell Biol 27:1264–1270

Shrivastav M, De Haro LP, Nickoloff JA (2008) Regulation of DNA double-strand break repair pathway choice. Cell Res 18:134–147

Shuck SC, Short EA, Turchi JJ (2008) Eukaryotic nucleotide excision repair: from understanding mechanisms to influencing biology. Cell Res 18:64–72

Skalka AM, Katz RA (2005) Retroviral DNA integration and the DNA damage response. Cell Death Differ 12(Suppl 1):971–978

Smit AF, Riggs AD (1996) Tiggers and DNA transposon fossils in the human genome. Proc Natl Acad Sci USA 93:1443–1448

Smith D, Zhong J, Matsuura M, Lambowitz AM, Belfort M (2005) Recruitment of host functions suggests a repair pathway for late steps in group II intron retrohoming. Genes Dev 19:2477–2487

Sonoda E, Takata M, Yamashita YM, Morrison C, Takeda S (2001) Homologous DNA recombination in vertebrate cells. Proc Natl Acad Sci USA 98:8388–8394

Stelter P, Ulrich HD (2003) Control of spontaneous and damage-induced mutagenesis by SUMO and ubiquitin conjugation. Nature 425:188–191

Sundararajan A, Lee BS, Garfinkel DJ (2003) The Rad27 (Fen-1) nuclease inhibits Ty1 mobility in Saccharomyces cerevisiae. Genetics 163:55–67

Symer DE et al (2002) Human L1 retrotransposition is associated with genetic instability in vivo. Cell 110:327–338

Szostak JW, Orr-Weaver TL, Rothstein RJ, Stahl FW (1983) The double-strand-break repair model for recombination. Cell 33:25–35

Takata M et al (1998) Homologous recombination and nonhomologous end-joining pathways of DNA double-strand break repair have overlapping roles in the maintenance of chromosomal integrity in vertebrate cells. EMBO J 17:5497–5508

Tower J, Karpen GH, Craig N, Spradling AC (1993) Preferential transposition of Drosophila P elements to nearby chromosomal sites. Genetics 133:347–359

Turlan C, Chandler M (2000) Playing second fiddle: second-strand processing and liberation of transposable elements from donor DNA. Trends Microbiol 8:268–274

Ulrich HD (2002) Degradation or maintenance: actions of the ubiquitin system on eukaryotic chromatin. Eukaryot Cell 1:1–10

Van de Lagemaat LN, Gagnier L, Medstrand P, Mager DL (2005) Genomic deletions and precise removal of transposable elements mediated by short identical DNA segments in primates. Genome Res 15:1243–1249

Van Luenen HG, Colloms SD, Plasterk RH (1994) The mechanism of transposition of Tc3 in C. elegans. Cell 79:293–301

Vanura K et al (2007) In vivo reinsertion of excised episomes by the V(D)J recombinase: a potential threat to genomic stability. PLoS Biol 5:e43

Walbot V, Warren C (1988) Regulation of Mu element copy number in maize lines with an active or inactive Mutator transposable element system. Mol Gen Genet 211:27–34

Walisko O, Izsvák Z, Szabo K, Kaufman CD, Herold S, Ivics Z (2006) Sleeping Beauty transposase modulates cell-cycle progression through interaction with Miz-1. Proc Natl Acad Sci USA 103:4062–4067

Walker EL, Robbins TP, Bureau TE, Kermicle J, Dellaporta SL (1995) Transposon-mediated chromosomal rearrangements and gene duplications in the formation of the maize R-r complex. EMBO J 14:2350–2363

Wang L, Heinlein M, Kunze R (1996) Methylation pattern of Activator transposase binding sites in maize endosperm. Plant Cell 8:747–758

Wang M, Wu W, Rosidi B, Zhang L, Wang H, Iliakis G (2006) PARP-1 and Ku compete for repair of DNA double strand breaks by distinct NHEJ pathways. Nucleic Acids Res 34:6170–6182

Warbrick E, Heatherington W, Lane DP, Glover DM (1998) PCNA binding proteins in Drosophila melanogaster: the analysis of a conserved PCNA binding domain. Nucleic Acids Res 26:3925–3932

Warren WD, Atkinson PW, O'Brochta DA (1994) The Hermes transposable element from the house fly, Musca domestica, is a short inverted repeat-type element of the hobo, Ac, and Tam3 (hAT) element family. Genet Res 64:87–97

Weia K, Kucherlapatib R, Edelmann W (2002) Mouse models for human DNA mismatch-repair gene defects. Trends Mol Med 8:346–353

Weil CF, Kunze R (2000) Transposition of maize Ac/Ds transposable elements in the yeast *Saccharomyces cerevisiae*. Nat Genet 26:187–190

Weil CF, Wessler SR (1993) Molecular evidence that chromosome breakage by Ds elements is caused by aberrant transposition. Plant Cell 5:515–522

Weinert BT, Min B, Rio DC (2005) P element excision and repair by nonhomologous end joining occurs in both G1 and G2 of the cell cycle. DNA Repair (Amst) 4:171–181

Weinstock DM, Richardson CA, Elliott B, Jasin M (2006) Modeling oncogenic transloca-tions: distinct roles for double-strand break repair pathways in translocation forma-tion in mammalian cells. DNA Repair (Amst) 5:1065–1074

Wessler S, Tarpley A, Purugganan M, Spell M, Okagaki R (1990) Filler DNA is associated with spontaneous deletions in maize. Proc Natl Acad Sci USA 87:8731–8735

Wu L, Hickson ID (2003) The Bloom's syndrome helicase suppresses crossing over during homologous recombination. Nature 426:870–874

Yamashita S, Mikami T, Kishima Y (1998) Tam3 in *Antirrhinum majus* is exceptional transposon in resistant to alteration by abortive gap repair: identification of nested transposons. Mol Gen Genet 259:468–474

Yamashita S, Takano-Shimizu T, Kitamura K, Mikami T, Kishima Y (1999) Resistance to gap repair of the transposon Tam3 in *Antirrhinum majus*: a role of the end regions. Genetics 153:1899–1908

Yan X, Martinez-Ferez IM, Kavchok S, Dooner HK (1999) Origination of Ds elements from Ac elements in maize: evidence for rare repair synthesis at the site of Ac excision. Genetics 152:1733–1740

Yandeau-Nelson MD, Zhou Q, Yao H, Xu X, Nikolau BJ, Schnable PS (2005) MuDR trans-posase increases the frequency of meiotic crossovers in the vicinity of a Mu insertion in the maize a1 gene. Genetics 169:917–929

Yoder K, Sarasin A, Kraemer K, McIlhatton M, Bushman F, Fishel R (2006) The DNA re-pair genes XPB and XPD defend cells from retroviral infection. Proc Natl Acad Sci USA 103:4622–4627

Yoder KE, Bushman FD (2000) Repair of gaps in retroviral DNA integration intermedi-ates. J Virol 74:11191–11200

Yu J, Marshall K, Yamaguchi M, Haber JE, Weil CF (2004) Microhomology-dependent end joining and repair of transposon-induced DNA hairpins by host factors in *Saccha-romyces cerevisiae*. Mol Cell Biol 24:1351–1364

Yu X, Gabriel A (1999) Patching broken chromosomes with extranuclear cellular DNA. Mol Cell 4:873–881

Zhang J, Zhang F, Peterson T (2006) Transposition of reversed Ac element ends generates novel chimeric genes in maize. PLoS Genet 2:e164

Zhou L, Mitra R, Atkinson PW, Hickman AB, Dyda F, Craig NL (2004) Transposition of hAT elements links transposable elements and V(D)J recombination. Nature 432:995–1001

Subject Index[1]

Page number followed by "*t*" indicates table.
Page number followed by "*f*" indicates figure.
"pp" indicates several following pages.

---

[1]Genes and their protein-products were not stringently differentiated by case sensitivity or italicization.